Contents

The Permaculture Book of DIY

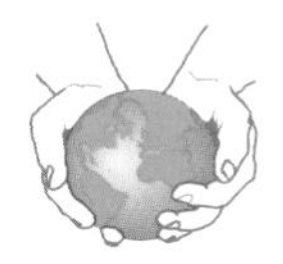

Permanent Publications

Published by
Permanent Publications
Hyden House Ltd
The Sustainability Centre
East Meon
Hampshire GU32 1HR
United Kingdom
Tel: +44 (0)1730 823 311
Fax: +44 (0)1730 823 322
Email: enquiries@permaculture.co.uk
Web: www.permanentpublications.co.uk

Distributed in the USA by
Chelsea Green Publishing Company, PO Box 428, White River Junction, VT 05001
www.chelseagreen.com

Designed and Typeset by Emma Postill

Printed in the UK by Cambrian Printers, Aberystwyth, Wales

All paper from FSC certified mixed sources

The Forest Stewardship Council (FSC) is a non-profit international organisation established to promote the responsible management of the world's forests. Products carrying the FSC label are independently certified to assure consumers that they come from forests that are managed to meet the social, economic and ecological needs of present and future generations.

British Library Cataloguing-in-Publication Data
A catalogue record for this book is available from the British Library

ISBN 978 1 85623 271 5

Introduction

What makes a DIY project 'permaculture'? Permaculture is a design framework with a set of ethics and principles modelled on Nature that helps us create low impact, ecologically sound projects. It is usually defined by gardens and land-based projects as they are often the easiest way of applying permaculture. However permaculture thinking can be also be applied to enterprise, farm, architectural and community design. At the other end of the scale, permaculture can be used to create far more achievable, self-build, practical DIY projects, large and small.

The key here is a number of over-arching ideas and principles:

- Design simple yet effective projects. Every project in this book has been well thought out and tried and tested.
- Make the project aesthetic as well as practical. Permaculture embraces design and form as well as function.
- Make every project replicable, both by providing clear instructions and photographs step by step, and by using techniques and tools that are accessible. Permaculture is all about sharing skills, information and resources, and being inclusive.
- Use a minimum of materials, preferably freely available within your community and recycled, repurposed or upcycled. Ensure those materials are as natural as possible. Avoid using composite wood, for example, or anything that requires a facemask whenever possible.
- Have fun building and consider sharing your skills with others during a build. Make it a community endeavour.
- Save lots of money by using reclaimed and unwanted materials. Make items instead of purchasing them and save even more money!

One of permaculture's aims is to work towards a regenerative culture, one that repairs in the broadest sense of the word and raises quality of life for all. In a world of experts and legislation, we are in danger of becoming less 'literate' with our hands, losing the confidence to design and build simple projects, and thus shopping for them instead. Yet all of us have a kid inside us that loves to build dens and 'play' outside.

Permaculture DIY is the root back to nourishing the kid in you whilst developing practical skills as an adult. Anyone, regardless of past experience and skills, can share this journey, find like-minds and mentors when we need help, and enhance our houses, gardens and community with the things we have made.

Maddy Harland is the editor and co-founder of *Permaculture* magazine. One of her great pleasures in life is to garden and build simple structures.

Enjoy Your DIY!

DIY is about taking time to be creative and there are a few guidelines that will keep you safe, especially if you are teaching a young person basic skills (many of these projects are perfect for imparting new skills under supervision). Firstly, work with clean, properly sharpened tools and in a clean workspace. Take your time and concentrate on the project in hand, allowing your mind the space to relax and stay focused. You will enjoy what you are doing far more. When we hurry and treat DIY like a chore we are more likely to have an accident. So treat it like a meditation! That's the permaculture way...

About the Authors

Mike Abbott – Champion the Lumber Horse
Mike has now retired from his woodland workshop but continues to make chairs for sale and to run chair-making courses in his workshops at home. His classic book *Green Woodwork* is now out of print but he has since written two other books, *Living Wood – From Buying a Woodland to Making a Chair* and *Going with the Grain – Making Chairs in the 21st Century.*

John Adams – Pallet Bench; Pallet Chair; Solar Food Dryer; Wicking Raised Bed; Triple Bay Compost Bin; Installing Small Scale Solar; Weave a Woollen Underblanket
John has recently retired from Permanent Publications and *Permaculture* magazine where he worked for over 15 years as creative director. He also turned his hand to a variety of practical DIY projects, from appropriate technoloy and eco-renovation to textiles and woodwork, documenting and writing articles about these projects for *PM*.

Stuart Anderson – Durable Pig Ark
Stuart and Gabrielle live on their 1.2 hectare (3 acre) permaculture smallholding in Brittany, France. They grow fruit, vegetables and firewood, raise sheep, pigs, chickens, ducks and bees and rent out their holiday cottage: **www.brittanycountrygite.com**

Katy Bryce and Adam Weismann – Making Homemade Paints
Katy and Adam run Clayworks (**www.clay-works.com**) and love working with clay wall finishes on a daily basis. They are also the authors of two books: *Building with Cob – A Step by Step Guide* and *Clay & Lime Renders, Plasters & Paints.*

Michel Daniek – Solar Electric Bikes
Michel, aka SolarMichel, was born in 1964 in Giessen, Germany. On reaching 30 he found himself totally dissatisfied with the German way of life, bought himself a truck and left in search of other ways to live. During his travels he experimented with a small solar system in his truck and has used solar energy in his day to day life ever since. In 1997 he finally settled in a new home in an alternative village in the sunny south of Spain. He is now married and father of two little daughters and works in many different ecological and sustainable projects all around Orgiva in Spain.

Lisa Gledhill – A Paper Pottery Kiln
Lisa is a film maker and writer working for conservation charity the National Trust. She's a keen allotment gardener and promoter of permaculture principles.

Steve Hanson – Heating Water with Compost
Steve is a permaculture teacher, practitioner and design consultant. A natural builder and professional craftsman for more than 20 years, he is co-owner of Permaculture Eden, a human scale regenerative permaculture farm near Lourdoueix-Saint-Michel in central France. **www.permacultureeden.com**

Ben Law – Bentwood Chair
Ben Law has always had a passion for healthy, biodiverse farms and woodlands. Having gained an Advanced National Certificate in Agriculture, Ben became a shepherd and set up a conservation landscaping business, specialising in ponds and wild flower meadows. Woodlands were a natural progression and, after seeking out a few experienced coppice workers, he began work in the woods and in associated coppice crafts. Ben was a founding member of the Forest Stewardship Council and has lived and worked at Prickly Nut Woods in West Sussex, UK, since 1991, training apprentices and running courses on sustainable woodland management, eco-building and permaculture design.
www.ben-law.co.uk

Simon Mitchell – Geodesic Growdome
Simonthescribe writes about permaculture, green living, politics and philosophy, sustainability, ecology and fun on his website: **www.simonmitchell.co.uk**

David Butler – Natural Swimming Pool
David is the director of BBC East 'Inside Out' programme and with his partner Alison and four children, Jasper, Theo, Felix and Otter are enthusiastic newcomers to permaculture. They live in Norfolk in an old barn with two acres and thirty chickens.
www.paganbutler.co.uk

Alicia Taylor and Jamie Ash – Rocket Stove Hot Tub
This hot tub was built as part of the Kents Collective permaculture project – a small community of surfers, running an organic smallholding and ecoretreat on the Cornish coast; experimenting with permaculture, communal living and self reliance. Kents Collective offers sustainable holidays in their yurts with home grown meals available. Alicia Taylor is one of the project leaders and a full time resident. Jamie Ash is a member of the collective, studying for an HNC in mechanical engineering.
www.kentsofcornwall.com

Beth Tilston and Will Harley – 'Bespoke' Wind Turbine
Beth Tilston and Will Harley live in a terraced house in West Sussex where they are working on becoming 'Partially Unplugged' from the grid. 'Partially Unplugged' is a design for living as sustainably as possible, without committing to major investment, or changes to a home, which may be outside the reach of many. A 12v system powered

by a single solar panel and a DIY wind turbine parallels the mains electricity. Home grown fruit and vegetables, and eggs from the chickens, supplement locally bought produce. Beth teaches scything (**www.learnscything.com**) as well as writing and taking photographs. Will is an analyst/project manager in the rail industry and enjoys general tinkering, cycling, and flying a vintage glider.

Mark Ridsdill Smith – Self-Watering Container Garden
Mark is founder of Vertical Veg, which inspires and supports food growing in containers. Visit **www.verticalveg.com** for container growing advice, seasonal tips and online training courses. Mark also runs the Vertical Veg Club (**www.theverticalvegclub.com**) – an online club for container growers all over world, to share experience, swap seeds, and learn from professionals in related areas, including permaculture, nutrition, wood work and horticulture.

Chris Southall – Simple Wood Fired Oven; Collecting and Cleaning Water
Chris and Rosie Southall are developing a life of suburban self-reliance in Clacton-on-Sea, Essex. They are a Permaculture Land demonstration project and WWOOF and HelpX hosts. You can visit **www.ecodiy.org** to follow their progress and find out more about the techniques they use to become more self-sufficient in energy and food.

Peter Willis – Cider Press and Scratter
Peter is occasionally inspired to design strange contraptions on the backs of envelopes, enjoying the challenge of reconciling the some-times conflicting requirements of simplicity, elegance and efficacy.

Tony Wrench – Reciprocal Framed Roofs
Tony has spent many years designing and implementing both renewable energy and building projects. He lives with his mate Jane Faith in the community at Brithdir Mawr, in Pembrokeshire, Wales. Their principles are sustainability, simplicity and spirit. Tony lives luxuriously, well below the poverty line, working on things permacultural and wooden. He makes his living from wood turning, singing and playing musical instruments (some homemade) with the local circle dance and Ceilidh band, Rasalila. Find out more at **www.thatroundhouse.info**. Tony is the author of *Building a Low Impact Roundhouse.*

Photo Credits

Photos and diagrams credited to chapter authors unless otherwise stated.

Choosing and Preparing Pallets

Choosing Pallets

Look for pallets that are clean and natural wood coloured. Avoid pallets with coloured planks or corners. Stains on pallets from cargo spills could be anything, so do not touch or use these. Unmarked pallets are made for national use and may or may not have been chemically treated. Look out for the following international codes on marked pallets:

- **HT** Heat Treated (no chemicals)
- **IPPC** International Plant Protection Convention (only HT allowed)
- **EPAL** Euro pallet (only HT allowed)
- **MB** Methyl Bromide fumigation (do not use or burn)
- **DB** De-barked

Any clean looking pallet carrying the IPPC or EPAL logos should be safe to use but use your own judgement.

Reclaiming Pallets

There are several ways to break up pallets to reclaim the timber, some of them easier than others. As we need to get the planks off with as little damage as possible I did most of the processing with an electric reciprocating saw fitted with a wood/metal blade. I soon learnt that using a combination of hammering from the back to loosen the joints and then using a spare plank as a lever, meant that I could cut through just the nails. You don't have to cut the joints mechanically; pallets can be dismantled manually using any combination of a short crowbar, claw hammer, pliers and hacksaw. If you do take them to pieces this way make sure you de-nail them as you go along, because if you don't sooner or later you will step or kneel on a nail with painful results.

▲ It helps if you loosen the joints first using a spare plank.

▲ Dismantling an old pallet.

◄ ▲ The pallet bench was inspired by these wonderful examples but had to be extensively adapted to work with standard euro pallets.

1

Pallet Bench

John Adams explains how to build a comfortable, two seated garden bench using only scrap pallets and a handful of screws

I am always interested in projects that use pallet wood and I have seen a lot of variations on the garden bench theme, some good, some bad, and some, well, downright ugly. The best I have come across for comfort and aesthetics are the seats at Malandelas[1] in Swaziland. Here they are used at the B&B (far left), outside the restaurant (left), for wedding functions and even in the House on Fire nightclub[2] – probably the only place in the world where you can regularly find royalty and pallet furniture together. They also hold the MTN Bushfire Festival[3] here where a lot more pallet furniture is on show courtesy of Guba[4] (see *PM*82), who seem to have real talent for it.

Having decided to have a go at making my own, I quickly realised some adaptation would be needed as the standard euro pallets commonly available here don't have those nice heavy bits of timber in them. I did keep the basic seat pitch and dimensions though.

Here's how I made mine but it's not the only way, so feel free to adapt.

Materials List

- 2 x 1,200 x 1,000mm pallets, plus a half one for spares
- 100 x screws 4 x 40mm
- 20 x screws 4 x 30mm
- 13 x screws 6 x 100mm

Construction Notes

To make a bench out of two pallets you might think that you just need to stand half of one on edge in the middle of the other to make an inverted T shape. There are several problems with this, but the main one is that the construction doesn't lend itself to easily joining them, without resorting to metal brackets or unsightly props.

My solution was to de-plank the base pallet and turn the ribs (the side and centre frame parts of the pallet) through 90°. This makes ribs that you can easily fix to, are stronger in the plane we want,

Author's update

The pallet bench is a great success and has proved to be very durable. The only thing I would change is the depth of the seat which is a little too deep unless back cushions are used. I would suggest reducing the number of seat planks by one. This will mean either repositioning the centre blocks in the ribs or cutting off the front ones and fixing them back into the shortened rib. If you do the latter you will need to adjust the leg length to achieve the desired seat angle (see step 2).

▲ 1 Remove the cross planks from first pallet. If you cut the last bottom one either side of the centre rib, you will end up with two temporary legs which are almost the right length to achieve the final 15° backward slope required. Flip ribs over to use as shown.

and look less pallet-like in the final piece.

There are various methods for de-planking pallets but I used a reciprocating power saw fitted with a metal/wood cutting blade to chop through the nails. Whether you do the same or use the traditional hammer and pry bar approach, it helps if you loosen the joints first using a spare plank (see page xi).

The second pallet will need deconstructing in the same way. I also needed a few bits off a third one as I ran out of planks before completion.

Hopefully, you can follow my step-by-step instructions. The only odd bit is those front legs. I wasn't sure how they were going to work out so I fitted them at step 9 but if you have prepared your materials in advance, logically they could be fitted much earlier.

When you have the back ribs fitted, as in step 8, you should have a structure which will take your weight and allow you to lean back on a rib without anything moving. If not, reinforce the frame as necessary.

For neatness I chamfered the ends of the leg bracing board and eventually sanded off the edges of the seat and back slats as well.

By the time you get to step 13 you should have a secure bench that you can sit on and have a well earned cup of tea, while you think about your final finishing.

References

1 **www.malandelas.com**
2 **www.house-on-fire.com**
3 **www.bush-fire.com**
4 **www.gubaswaziland.org**

▲ 2 On a level surface, set up a rib and trim its temporary leg to about 320mm so that top slopes at approx 15°. Copy onto other rib. Don't worry about being super accurate as long as all the parts match. I built it by eye; the level and tape are only there for the photos.

▲ 3 Space out the three ribs so that a cross plank just overhangs the sides. Join the ribs together with a plank either end. Lay out the other five seat planks and screw down the last one just in front of the centre blocks. You could jump to step 9 and fit the front legs at this point.

▲ 4 Now for the second pallet: De-plank it until you are left with just the three ribs. Cut these just above their centre blocks and you now have the three ribs for the back. A quick look at step 8 may help to explain what we are trying to achieve.

▲ 5 Prop up one of the back ribs so it rests as shown, i.e. with the leading edge in line with the top of the last seat plank. It should make about the right back angle (115° approx). Trim it off until it stands properly and then copy to the other two ribs.

▲ 6 Put the back frame ribs in place and using three long screws per rib, fix down by screwing through the lower block into the block in the frame below. The ribs' original construction may not be as strong as it needs to be, so add a few screws to strengthen as needed.

▲ 7 Tip bench on its side and screw a cross plank along the bottom of the three base ribs. Screw it into the centre blocks and each rib member. Now using off-cuts, fix strapping pieces between the back ribs and seat base ribs. Cut the ends of these to match the back angle.

▲ 8 One reinforcing strap goes on the inside of each outer back rib and one either side of the centre one. You will need five of the short screws for each strap into the back ribs plus three longer ones into the block. If anything still wobbles reinforce with extra screws.

▲ 9 Remove the existing legs and use one as a template to make two copies for each leg. Sandwich together and trial fit to an inner rib member as shown. The middle piece will be displaced by the depth of the rib. Mark and trim as required.

▲ **10** Screw the pieces together from both sides and fit to the ribs using wood block spacers. Fix by using three long screws from either side. Add a plank across the front face of the legs, this not only helps brace the legs but acts as a support to the centre rib.

▲ **11** It only remains to fix down the seat planks and slat the back. Attach all the seat planks with screws into every rib section. Stagger the screws to lessen the risk of splitting. Fix the bottom slat on the back ribs leaving a gap for drainage. Now fix the top slat.

▲ **12** Find the centre between the top and bottom slats and fix the middle one there. Fill in the remaining two slats and that's it, job done.

▲ **13** The completed pallet bench ready for sanding and painting.

2

Pallet Chair

John Adams shows you how to make a comfortable Adirondack style outdoor chair from reclaimed pallet timber

I have long been fascinated by the American style of outdoor chairs, usually called Adirondack or Plank chairs. They are to be found on porches, in backyards, by lakes and waterfronts over a large part of the USA, and often feature in the foreground of pictures, like those stunning views of New Hampshire and Maine you find on calendars.

I thought with a little research I could probably make one myself, then I realised that if the timber proportions were adjusted, it could be made for minimal cost out of timber reclaimed from old pallets and a handful of stainless steel self drilling screws. So I set about trying to do this – and you can too – with no more than a quick sketch on a piece of scrap paper. I calculated the chair would need about 24 pallet slats. I chose to use various widths for mainly aesthetic reasons, i.e. a wider panel in the centre of the back with narrowing planks to the edges, so it wouldn't look too 'pallet' like.

◄ The completed pallet chair before being painted.

Acquiring the Pallets

I was lucky that I already had a pile of redundant pallets just outside the office, but they are not normally hard to obtain once you start looking. Damaged and sometimes undamaged ones can often be found in skips or piled up on building sites. Usually they are free for the taking, but do ask first. For a guide to choosing and dismantling pallets see page xi.

The Chair

I largely made the chair up as I went along, adapting the outline design to the materials available. The result doesn't look exactly like the original sketch, the main change being that both the seat and back are now solid rather than slatted because that's the way the plank widths worked out. The following instructions will allow you to build a replica, but feel free to adapt and alter it to suit your materials and preferences.

Materials List

- 3 old pallets (1 large 1 x 1.2m and 2 medium 0.8 x 1.2m)
- Screwfix[1] Turbo Ultra Stainless screws (4 x 45mm and 4 x 35mm)

Tools

- Basic hand woodworking tools and a crosshead screwdriver, though using power tools will speed the job up.

Cutting List

Frame

- Back legs (130mm) – 4 x 1,000mm
- Front legs (100mm) – 4 x 560mm
- Front cross-member (100mm) – 1 x 560mm
- Inner rear cross-member (100mm) – 1 x 500mm
- Outer rear cross-member (100mm) – 1 x 495mm

Seat

- Slats (80mm) – 6 x 560mm widest – 535mm narrowest

Back

- Centre (130mm) – 1 x 1,200mm
- Sides (100mm) – 1 x 1,200mm
- Arms cross-member (100mm) – 1 x 620mm
- Top cross-member (100mm) – 1 x 480mm

Arms

- Underarm braces (100mm) – 2 x 600mm
- Arm tops (145mm) – 2 x 730mm

▲ 1 Template for back legs.

▲ 2 Screwing together a T-piece front leg.

▲ 3 Top face of front cross-member showing chamfer and cut out in the inner back leg.

Assembly

Read through the instructions and study the pictures before beginning. The dimensions on the cutting list are a guide only and you may want to adjust them to your chair as you build it.

Back legs

These are the legs that run from half way up the front legs to the ground at the back. Take one piece and from a square end mark 50mm up and 160mm along, join with a line and cut off, this will be a back foot. Mark 900mm along the top surface and then nip or round off the corner. Stand the piece up on its foot on a level surface. Using a spirit level mark down the face from the 900mm mark to establish the front angle and cut. Mark this piece TEMPLATE and use it to make the other three (1).

Front legs

The front legs are made as offset T-pieces. Take a piece and draw a line down the length 20mm in. Butt another piece up to this line to form an offset T. Clamp in place and screw together from the front with four 45mm screws (2). Now viewed from the front, the longer arm of the T should be to the right of the right hand leg. Mark this arm 100mm from the top, and at the bottom mark a point 60mm in from the left hand edge. Connect line and cut off. Repeat to make the other leg (don't forget to check you are making a handed one).

Cross members

The first cross member goes just inside the front legs and is inset into the back ones, this needs to be chamfered off along its top edge at a roughly 20° angle (3). The other two support the base of the chair back.

Frame sub-assembly

Take one of the inner back legs, sub-assemble to the inside of one of the front T-piece legs about 345mm up and clamp. Offer up the front cross member with the chamfer uppermost and mark where it needs to be cut out of the front of the inner leg. Cut out and copy with opposite inner back leg. Insert front brace and screw into position with 45mm screws. Attach back leg to front with four 35mm screws, plus a 45mm screw through from the front. Repeat to make second side.

Use a square to mark lines at 90° across the inner faces of the back legs 480mm back from the front edge. Fit the longer rear cross member inboard of this line and fix with 45mm screws. Fix the second cross-member 25mm aft of the first one (4).

To finish the frame, fix a couple of bits of scrap as spacers to the outside of the inner back legs with 35mm screws. Attach the outer back legs using 45mm screws.

Author's update

This chair is a popular addition to my garden and has survived very well, particularly considering it was untreated for several years. The only changes I would make are to leave a gap between the seat base and the back, to make it drain better and maybe to recline the back by a few degrees. Try the angle out before final fixing.

▲ 4 Frame sub-assembly with front and rear cross-members, and first seat slat in position.

▲ 5 Seat slats.

▲ 6 Single screw holding centre back plank.

▲ 7 Notched outer back plank.

▲ 8 Underarm brace.

▲ 9 Making the arm tops.

Seat slats

The seat extends back 480mm from the front edge. The seat tapers – front to back so each of the six slats is 5mm narrower than the previous one. They are fixed with four 35mm screws per slat (5).

Seat back

Take the central piece for the back and place it in the middle between the rear cross members. Fix with a single central 45mm screw (6). Square up and fix with a further four 45mm screws and a couple through the inner rear cross-member. Fit the remaining planks either side in a similar fashion. The outer planks will probably need to be notched to fit (7).

Prop the unsupported back up so that it is at 90° to the seat base, then using a spirit level from the top of the front legs mark the level position on each side of the seat back. Mark off 40mm from either end of the arm cross member and cut away the tops to make flat landing points for the arms. Making sure the flats line up with the level marks, fix with 35mm screws to each plank.

Cut underarm braces to the angle of the back and fix to inner T with 35mm screws and into the back with 45mm, also screw through the cross member (8).

▲ **10** Chair prior to cutting back curve.

The arms

Cut a 40 x 40mm notch on the top rear surface, it needs to be angled, so that the arm will fit snugly into the chair back. Measure 40mm outboard of the notch and mark. Find a second point 200mm back from the front on the outside edge. Join these two and cut to form a tapered arm. Use as a template to make the other arm (remember to flip it over so they are handed) (9). Sand to round off the main edges and fit into position. Fix down each arm with 45mm screws.

Finishing off

Try your chair out for size (10) and once you are lounging comfortably take a pencil and mark the back just above your head. Find a large circular object like a dustbin lid and draw round it to mark an arc on the seat back. Cut along this line to form a curved seat back. Fit the top cross-member with 35mm screws and you're done. Sand any raw edges, stand back and admire your chair.

References

1 Screwfix: **www.screwfix.com**

3

Solar Food Dryer

John Adams describes how he made a solar food dehydrator and how it performed in a less than typical British summer

▲ The finished products.

◀ The finished solar dryer loaded with a range of fresh produce.

In many parts of the world solar drying is a common way of preserving food. We are all familiar with dried fruits for cake making, dried banana crisps, and sun dried tomatoes but dehydration is not a preservation technique that seems to have caught on in the UK and Northern Europe.

As far as I can see, there is no reason why it shouldn't work in a cool temperate climate, as all that is required is two sunny days in a row with temperatures of 10°C or above. We should be able to manage that even in the worst British summer and with the recent increasingly warm late summers, well into the autumn as well (readers in warm temperate and subtropical climes should have no problems).

To see if it would work I decided to build and test a proven American-designed solar food dryer using plans from the excellent book *The Solar Food Dryer*.

Materials List

Timber

- Sides, base and rack frames, 2.44 x 1.22m sheet of 12mm exterior plywood
- Legs and cross brace, 3m of 44 x 44mm softwood
- Internal rails, approx 6m of 18 x 18mm softwood

Metal

- Sheet steel 788 x 500mm
- Aluminium channel for glass, 3m

Glazing

- 4mm clear glass, 763 x 695mm

Screen Material

- 2 x 1.2m of charcoal fibreglass material with a square mesh hole size approx 1.6mm

Hardware

- Screws size 6:
  - Countersunk – 50 x 32mm, 25 x 25mm, 25 x 18mm, 25 x 12mm
  - Roundhead – 25 x 12mm
- 2 x Small corner brackets
- 6 x Glazing brackets
- 2 x Small plain hinges
- 2 x Small strap hinges
- 1 x Friction stay
- 2 x Small brass handles
- 2 x Carrying handles
- 1 x Glazing Seal 3m of foam weatherstrip
- 1 x Tube clear silicon sealant

Heater

- 120W tube heater 660 x 52mm

▲ 1 Plywood marked up for cutting.

▲ 2 Cut components, the panels at the back will become racks.

▲ 3 Sub assembly showing legs and rack supports.

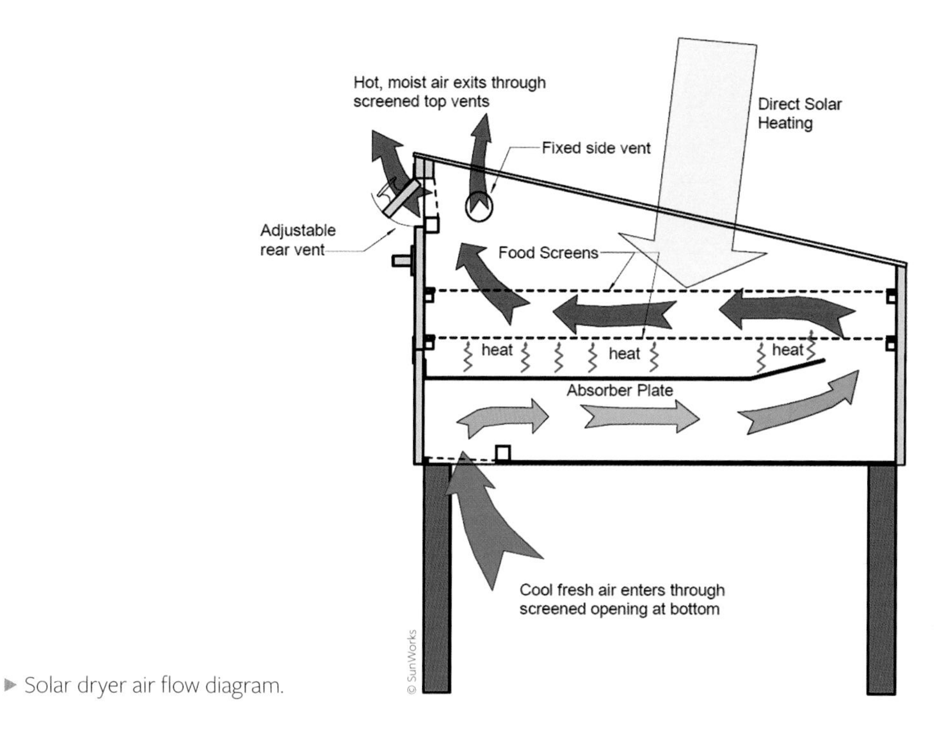

▸ Solar dryer air flow diagram.

The dryer is basically a wooden box on short legs with a black metal solar absorption plate, two drying racks and a glass top. Sounds simple, but the secret of its success is the carefully thought-out airflow through the unit (see diagram above).

Building The Dryer

The original plans are in imperial units and required a little adjustment to work economically with metric materials; the lack of suitable aluminium framed screens at affordable prices forced some redesign, but other than that it was built by the book.

Cutting the main components

The sides, base and rack frames were all cut from a single sheet of 12mm exterior plywood (1). I drew it all out on the sheet first, except the base which was made from two offcuts, and then cut out the major components (2).

The rack frames were put aside to finish off at a later stage.

Next I cut the softwood legs and internal rails. The chamfered brace rail which goes at the back is specified as being cut out of the leg material, which I did, but in retrospect I think it would have been easier to find a piece of wood nearer the correct size in the first place.

Having cut all the main parts it is tempting to fix the rails in position on the sides, fit the legs and assemble the main body of the solar dryer, which I did (3). What I should have done, however, was address the issue of the solar absorber plate first.

▲ **4** Metal bending rig.

▲ **5** Sprayed absorber plate.

▲ **6** Tube heater in front of unit.

▲ **7** Aluminium edging fixed to the glass with silicone sealant.

▲ **8** Ready for glazing, showing the glazing brackets and sealing foam.

▲ **9** A drying rack ready to load.

Solar absorber plate

The solar absorber is made from a piece of steel or aluminium of a bendable thickness. It needs cutting to fit, including cut outs for the rear legs. The front edge also needs bending up which I achieved by clamping it between two stout pieces of wood where I wanted the bend to be (4). The plans also call for the side edges and back edge to be folded up by 25mm to be used as fixing brackets to the sides. I ignored this as being too complex in steel and just cut it to fit, with oversize holes (to allow for expansion) drilled through the plate for fixing down onto the support battens. Check it will fit, degrease the metal and spray with a couple of coats of stove black (5). I did all of the above at a later stage and had to partially dissemble my drier to fit it.

Electrical backup

If you want to remove the uncertainty of getting two days of suitable conditions or to extend the drying season, you can fit electrical backup. The plans call for two x 200W incandescent light bulbs fitted below the absorber plate. This seems like a lot of watts for such a small space so I fitted a 660mm, 120W tube heater (this didn't produce enough heat to dry on its own but did raise the temperature by about 16°C and helped keep humidity down) (6). If you want to go for either of these options it is probably easier to do it before fitting the absorber plate.

Glazing

The top of the dryer only needs to be single glazed, though the glass does need to be tough enough to make it safe. I used 4mm which I edged with aluminium channel. The edging was mitred at the corners and fixed to the glass with clear silicon sealant (7). The glazing panel was then fixed onto the box, which had self adhesive draught excluder tape on it, with glazing clips made from cut down little L brackets (8).

Drying racks

The drying racks should be made of aluminium framed fibreglass or poly-propylene screen material. However as getting these made up was prohibitively expensive, I went for cutting the frames out of the ply sheet and covering them with fibre-glass screening which I had sourced by the metre (see Resources) (9).

The cut out frames were only about 20mm wide. Their rigidity is based on having curved internal corners (I drew round a cup).

Once cut out and sanded the frames were covered with tightly stretched screening material, stapled in place along the edges.

The offcuts of screen material were then used to cover the vents in the rear, sides and bottom of the dryer.

▲ **10** Probe thermometer fitted into the case.

▲ The completed solar dryer.

▲ **11** Loading racks through rear door.

© SunWorks

© SunWorks

▲ Slices of pear before and after solar dehydration.

Controls

The aim is to keep the temperature in the dryer at between 48°C and 66°C. Too cool and the food may spoil, too hot and it will cook instead of drying.

To monitor the temperature fit a probe thermometer into the case (10).

I used one from an old smoker.

The controls are pretty simple, basically there is an adjustable vent at the back which needs to be opened or closed accordingly. Also for optimum performance the dryer too can be moved, using the two lifting handles, to track the sun or not as required.

Recycled materials

Most, or possibly all, of the solar dryer could have been made from recycled materials. In fact I had intended to make most of it from thin pallet planks but lack of time and readily available resources forced me down the new materials route.

If you have time it would be well worth collecting at least the major components. The three most expensive items were the sheet of exterior ply, the metal sheet and the glass. The hardware largely consists of handles, knobs and hinges which should be easy enough to come by, which only leaves the aluminium frame for the glass and the screen material to buy.

Drying in the UK

The inclement weather we suffered in June did nothing to help but I did eventually find a couple of consecutive days which were forecast to be at least partially sunny.

I wanted to dehydrate a cross section of produce you might end up with a glut of, so I used apples, tomatoes, mushrooms, peppers and chillies. These were sliced to be no thicker than 6mm, laid out on the racks (they can take about 1.25kg each) and loaded into the drier (11).

The end of day one saw encouraging progress with noticeable drying. I switched on the heater overnight, which kept the temperature at around 27°C. Day two, frustratingly close, the thin produce like chillies were perfect but the thicker, wetter stuff just needed a bit more drying. Day three, and a few more hours of sun finished them all off nicely.

So does it work in the worst a British summer can throw at it? Yes it does, but it may need three days, not two.

If you would like to build your own solar dryer, I recommend the modest investment of buying the book which also contains a wealth of additional information to help you make a success of your food dehydration.

Resources

Timber, metal and hardware

B&Q:
www.diy.com

Screen material

Flyscreen Queen:
www.flyscreenqueen.co.uk/flyscreen-material/charcoal-flyscreen.html

Further Information

SunWorks Technologies LLC:
www.solarfooddryer.com

4

Wicking Raised Bed

John Adams explains how to build a self-watering raised bed for free – well almost – using pallets and other scavenged materials

Materials List

- 2 x 1,200 x 1,000mm pallets – other sizes will work
- 4 x planks for seat edges – these can be pallet planks but wider boards are better
- 60 x deck screws 4 x 50mm
- Assorted short screws, nails or staples
- Some off-cuts of plastic drain pipe
- Lining material (e.g. sign boards)
- Floor (optional)
- Waterproof membrane approx. 2,000 x 2,000mm
- Wicking material (e.g. coffee sacks)
- Reservoir material (e.g. gravel)
- Soil (top soil mixed with compost)

The Food is Free Project in Texas,[1] featured in *Permaculture* magazine issue 81, encourages the making of self-watering raised beds from discarded pallets and other scrap materials. I was intrigued by these apparently simple but very effective raised beds that use wicking from a reservoir to keep them moist and productive for up to two weeks without outside intervention. As no step-by-step plans seemed to exist I decided to build one of my own, and record and share the experience.

This is how I built a 1,200 x 1,200 x 500mm bed.

▶ The Food is Free planter which inspired the project.

◀ Author with the finished wicking pallet bed.

▲ 1 Source two clean, strong pallets.

▲ 2 Remove excess planks with a crowbar and de-nail.

Building the Frame

I already had a collection of pallets scrounged from our delivery drivers so I only had to select some strong clean ones for this project (1). I chose to use two large ones to make a square bed but you can mix and match to make beds of other sizes and shapes.

The intended height of the finished bed was about 500mm which meant the pallets only needed to be cut in half. You can make them lower than this if it suits your pallets better, and they will require less material to fill.

Both pallets had a central foot which needed removing with a crowbar. One of them had a gap between planks in the centre but the other had a plank which required careful removal using a metal cutting blade in a reciprocating power saw (you could use a crowbar and nail puller instead) (2). The removed feet and plank were kept for later use.

Both pallets were cut in half with the saw (3) and the pieces laid out on a hard surface facing each other in pairs to form a square.

At this stage I realised they were not going to join together very easily so I flipped over two facing pieces and tried again (4). This was much better as they could now be locked together as each corner had a block top and bottom to screw into (5). I put three deck screws through the planks into each corner block (6).

The result was a pretty strong box but one with a wide top on two sides and a thin edge on the other two. I then fitted a seat edge around the top using two of the leftover pallet planks and some wider boarding I found behind my shed (7).

While looking for the planks I also found some 12mm ply and decided to use it to make a floor (this is entirely optional and is not normally done).

▲ 3 Cut pallets in half using a power saw fitted with a wood and metal cutting blade.

▲ 4 Make a square by laying flipped halves out on a hard level surface.

▲ 5 Corner fitting detail.

▲ 6 Fix the sections together by screwing into the blocks at the corners with long screws.

▲ 7 Cut seat planks for the tops.

▲ 8 Move to site and level.

The Floor & Lining Out

To support the floor (it is going to have to take a lot of weight), I screwed the last foot plank and blocks in the centre of the bottom of the box. The floor was then cut out of the sheet of ply as a single piece. It was only then that I realised it wouldn't fit without taking the seat edges off again. So if you go down the flooring route, fit the seat edges afterwards. It also occurred to me that it was going to be difficult to relocate once the floor was in, so I moved the frame to its final position and with a bit of digging got it reasonably levelled (8). That done, the ply floor went in and was firmly screwed down and the seat edges refitted.

The Food is Free Project recommends lining the frame on the inside with political campaign boards (estate agents' signs would also work), but I lined mine with thin plywood which came as protective wrapping on a furniture delivery. This was lightly screwed in place wherever I could get a fixing (9). If you don't have a floor, line the ground with boards or carpet to protect the waterproof membrane.

The Waterproof Membrane

The inside of the bed needs a waterproof lining and luckily I still had some offcuts of butyl rubber from my last pond. If you're not so lucky, any strong plastic sheet or tarp will do providing it holds water. Not so luckily, my largest offcut was just under size. It fitted the bottom but only came half way up the sides. As this was still deeper than needed for the reservoir, I attached it to the sides with a staple gun (only along its top edge of course) and then hung other offcuts to overlap from above (10). These were held along the top by pieces of screwed-on batten. Normally the lining would be one piece, and secured by wrapping around the battens, which are then fixed just below the seat edges.

Reservoir & Wicking Membrane

The design calls for a reservoir about 150mm deep, composed of gravel, recycled glass beads or other small aggregate, with a wicking layer laid on top. The wicking layer can be any absorbent material, old bedding, curtains, horticultural fleece etc., but old coffee sacks are recommended. I can see why as the double thickness hessian has obvious strength and wicking potential. This got me thinking about wicking and I couldn't see how the wick was going to work when it is only in contact with the top of the reservoir, so I decided to modify the design slightly. Having sourced some hessian coffee sacks on eBay[2] for a modest amount (the only bought item so far), I part filled four of them with old aggregate (11). Laid in the reservoir and tamped down to level, these allow wicking from the very bottom.

Next came a bit of plumbing. I used a piece of guttering down pipe for the filler. This was cut to the inner depth and fixed in a corner in such a way as too leave a gap at the bottom (12). The overflow was made from a piece of 50mm plastic drain pipe. Large holes were drilled along what would be its bottom edge and it was laid on top of the bags of aggregate. The position was marked on the side wall and a hole cut through (13). Once fitted, holes downward, the whole lot was covered with two more coffee sacks (14).

▲ 9 Fit floor (if used) or ground protection, line sides with thin boarding.

▲ 10 Lay waterproof membrane and fix with battens.

▲ 11 Put in a layer of ballast, in this case old gravel in coffee sacks.

▲ 12 Fit filler pipe with a space underneath.

▲ 13 Drill through side and fit a perforated pipe for the overflow.

▲ 14 Add a layer of hessian.

▲ 15 Fill with soil and compost and mix well to create a planting medium.

This seemed like a good time to test the bed before finishing it off. I decided to fill it with water up to the overflow level. Just before I filled it, I got cold feet about the strength of my floor and loose fitted a couple of bits of 4x2 under it. When filled with a hose I was delighted to find that the only problem was that the overflow pipe needed lifting slightly at its free end.

Soil Layer & Planting Up

All that remained was to add the soil layer and plant it up. Even with the aggregate added, there was still quite a daunting volume left to fill – approximately 450 litres. I harvested as much compost as I could from my Hotbin and scrounged around for other suitable material. Unfortunately having recently completed another bed I could only come up with about a third of the volume and had to purchase the rest. I used roughly equal quantities of well rotted horse manure, composted organic matter and top soil. These, along with my compost, were added to the bed in layers. My daughter then mixed it all together by hand (we tried using a fork but found it caught in the hessian) (15).

It just remained to plant it up with plugs. We put in a range of vegetables and clumps of herbs in the corners.

Only time will tell how well it works, but so far I am really pleased with it. It seems well worth the modest number of working hours it took to complete and I look forward to eating the produce.

If you decide to follow my design the only things I would do differently are to fit the two extra supports under the floor at the construction stage and allow more time to get the fill materials together. If you do that, you should end up with a really strong, low maintenance, raised veggie bed – quite possibly for free.

References

1 **www.foodisfreeproject.org**

2 **www.ebay.co.uk/usr/dumdumbig**

▲ The completed bed planted up with vegetable plugs and herbs.

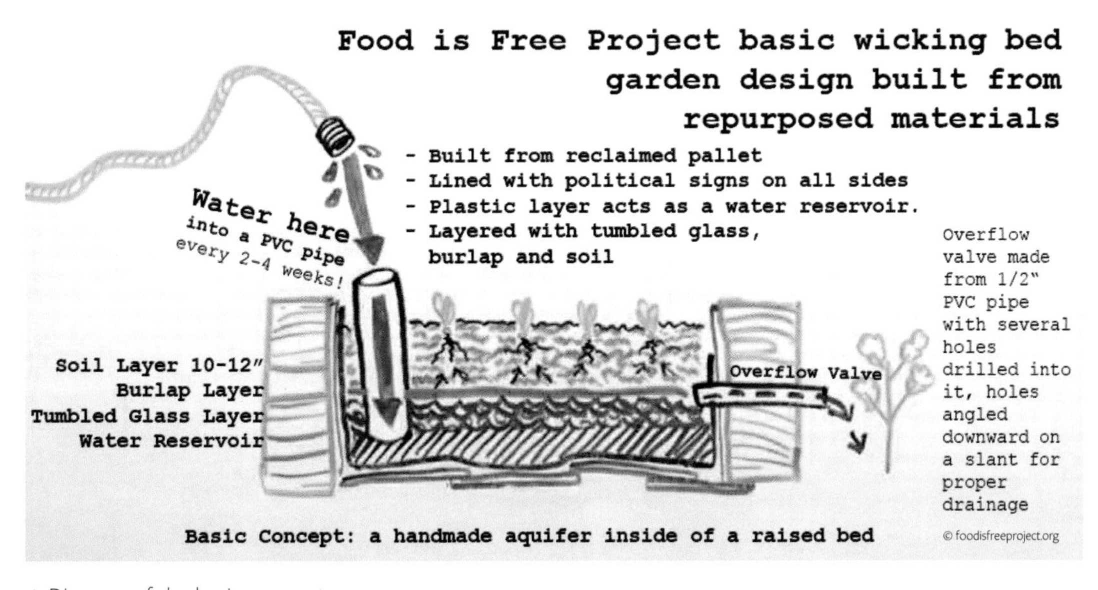

▲ Diagram of the basic concept.

THE ASSOCIATION OF
LATHE TURNERS

5

Champion The Lumber Horse

Mike Abbott describes how to make a modern shaving horse from recycled wood, so you can ride off into the sunset and get green woodworking!

After thirty years or more of using shaving horses, the occasion arose to have a total rethink of their design. I had always based my shaving horses on a 1.2m length of log, about 30cm in diameter, but I was aware this is not the sort of thing that most people have lying around the workshop. For some time I had wanted to come up with a design that used easily obtainable materials – for people without access to woodlands. While recently erecting a new workshop I discovered the effectiveness of using cordless drills to drive modern coachscrews into softwood beams. I had also spent a few hours that summer chatting to Owen Jones, a swill-basketmaker, while he was sitting astride his shaving horse designed for gripping thin slivers of oak. It had a central arm slotted through a horizontal platform and I had been interested in exploring this design.

◄ The finished lumber horse, made from easily obtainable sawn softwod beams, is an essential tool for green woodworking projects.

Despite my lifelong mission to persuade people of the advantages of cleft, unseasoned hardwoods, I ended up with a design made out sawn softwood beams. It needed a name and when I used the term 'lumber horse', it rung a bell from my childhood TV viewing – Champion the Wonder Horse – and the name 'Champion the Lumber Horse' has stuck.

It can be made in less than a couple of hours by almost anybody, as can be seen by the photos taken at a session of our youth club in our local village, Bishops Frome.

In future, I intend to use locally grown Western Red Cedar, which should work as well as, if not better than, the stock from the timber yard. My assistant Tom assures me that a ten minute sortie through the skips along most urban streets would easily yield sufficient raw materials for the job. There are no precise joints needed and the only woodworking skills involved are the ability to wield a handsaw and a drill. I suggest you use the sequence illustrated but there's no reason why you shouldn't assemble it any way that takes your fancy.

Materials List

- 4 x 2.4m lengths of sawn, seasoned 100 x 40mm softwood
- 50cm length of roofing batten
- 40cm length of hazel rod
- 35 M6 turbo coach screws, 90mm long (although ordinary screws or nails would be possible but far less fun)

Tools

- An electric drill, mains or cordless (or a brace or a bar auger)
- 25mm drill bit
- 8mm hex nut driver (or a hammer if using nails)
- Handsaw
- Tape measure
- Pencil
- An accomplice (Roy Rogers) to hold things, or a good cramp
- A couple of low benches

Pre-drilling the holes

You are going to need several 25mm holes in the platform and arms of the horse. If you have the use of a pillar drill and/or a bench vice, you will probably find it easier to drill the holes before assembly, in which case drill two holes in each section of the platform and three in each arm as illustrated. You may rather drill these holes once the horse has been assembled, which is fairly easily done with an electric drill, a brace and bit or a bar auger.

Notes

- An alternative is to use a single arm (see illustration below right), best held together with strong bolts partly for strength and also so that it can be taken to bits to fit into the platform. The arm will also need to be planed a few millimetres thinner to move freely.

- Another alternative is to pivot the arms with a metal pin or large coach screws, in which case the holes in the platform and the arms should be smaller than 25mm.

- If you need to transport the horse, the rear leg assembly can be easily removed by simply removing the few screws that hold the tops of the legs to the seat and the leg brace to the rear spacer.

- You could trim the angles off the tops and bottoms of the legs but the tops of the back legs can be very handy as a bench-stop when the horse is in use.

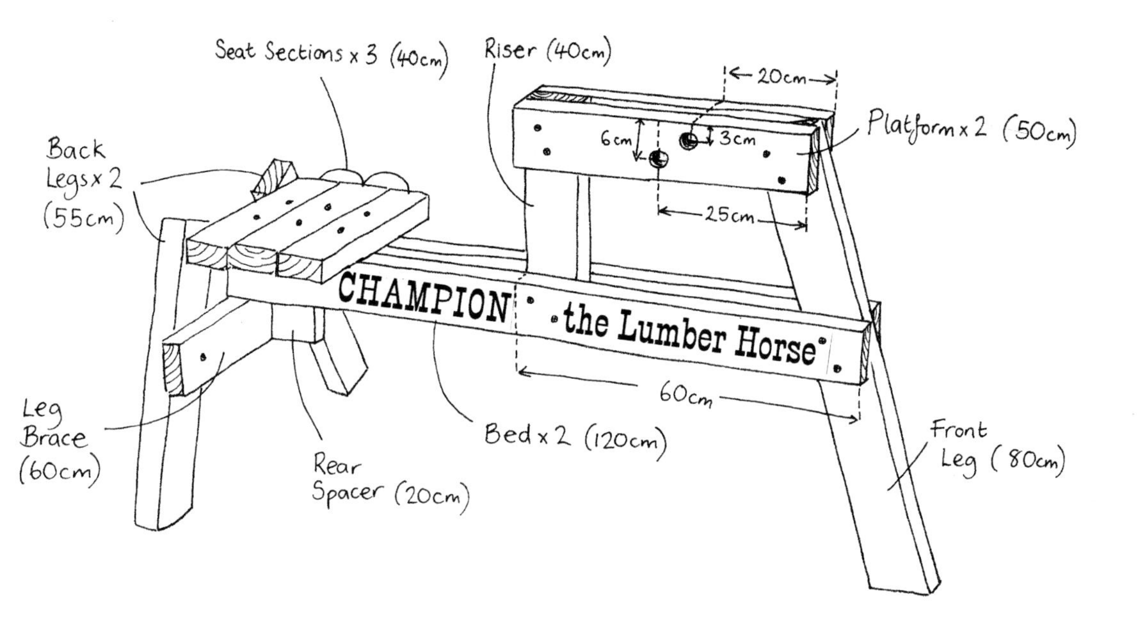

▲ Main lumber horse frame construction diagram.

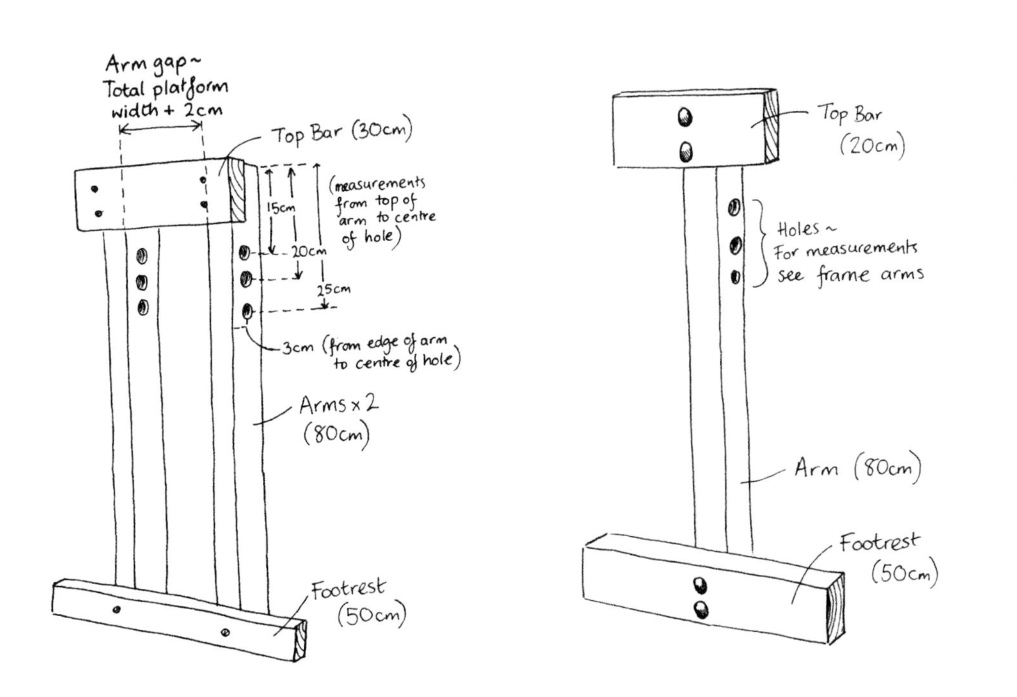

▲ Standard twin swinging arm diagram.

▲ Alternative single swinging arm diagram.

▲ 1 Fix one section of the bed to the front leg, the central riser and the back spacer, using just one screw at each joint.

▲ 2 Fix one of the platform sections to the tops of the front leg and the riser so that it is about parallel with the body section. The holes in the platform should be positioned as illustrated.

▲ 3 Turn the whole assembly over and lay it down with the riser about square to the bed and with the front leg sloping. Fix the other sides of the bed and the platform with a couple of screws at each joint. Take care that the screws are not right at the end of any components, as they would be likely to split the wood.

▲ 4 Fix the rear seat component to the back of the body.

▲ 5 Fix the leg brace to the rear spacer against the underside of the body.

▲ 6 Stand the horse up in the air onto the two back legs with the tops of the legs just protruding above the seat and fix the leg brace to the legs with one screw at each joint.

▲ 7 Stand the horse on its feet and fix the top of each leg to the end of the seat component. Then making sure the legs are splaying equally, drive another screw through each leg into the leg brace.

▲ 8 Fix the remaining seat components into place.

▲ **9** Screw the top bar onto the arms so that the gap between the arms is about a couple of centimetres wider than the width of the platform. Now fix the footrest (which could be another 50cm length of 100 x 40mm, a strong length of 50 x 25mm or anything in between).

▲ **10** Lift up the front of the horse, slide the frame into place and pivot it with a short hazel rod, a length of dowelling, a length of broom-handle or a specially made 21mm wooden pin. This should be tapered at one end to make it easier to poke in when adjusting the gap.

▲ And now you can ride off into the sunset, ready to make a whole host of wooden artefacts.

The Chestnut Pole Shave Horse by David Watson

As a professional coppice and green woodworker I have inevitably accumulated a vast array of tools and equipment to make my working life easier, faster and more efficient. Amongst this collection of equipment is a small group of tools that I consider essential; these include bill hooks, drawknives, froes and axes and of course, my humble shave horse. Together with my treasured Gilpin drawknife, the shave horse is used on an almost daily basis for peeling and shaping; it's fair to say that I'd be lost without it. I started out using simple shave horses that consisted of a cleft chestnut log upon three legs and a simple hinged wooden board or plank and chock arrangement that worked okay but I quickly found that its gripping capabilities were limited and the workpiece would often slip from the grip of the horse. As a one off occurrence this isn't a problem but if you're peeling several hundred cleft pieces of chestnut for an order of gate hurdles or dressing a few thousand shakes or shingles, then this can quickly become frustrating. The need for a rethink had arisen. The resulting shave horse couldn't look anything further from the likes of 'Champion the Lumber Horse' but using it has been nothing short of a pleasure. It utilises an upside down 'Y' section of Chestnut pole, cleft in two, and a series of holes and pegs that mean its fully adjustable to take large diameter poles right down to shingles just a few mm thick without slippage. Together with my drawknife, I use the shave horse regularly, from peeled poles to making ornate gates, benches, chairs and sheep hurdles. I take great pleasure in sitting down at my horse to peel the horizontal rails for gates and hurdles before shaping the tenons on either end with my drawknife.

▲ (top): Finished bins with cedar shingle cladding.
▲ (above): Surplus cedar shingles from the roof of another build were used to clad the compost bin.
◄ Compost bin in use following construction.

6

Triple Bay Compost Bin

John Adams describes how Mark Fisher made his stylish triple bay compost bin out of old pallets and left over materials and invites you to make one too

Part of Mark Fishers brief for re-landscaping the Ecology Building Society's new headquarters, (as featured in *Permaculture* magazine issue 48), was the construction of a large compost bin. We were so impressed with the finished item we asked Mark how he made it so we could show you how you could make one too.

The basic design is for a three bay New Zealand style composter constructed from old pallets. Mark beautified his by cladding it with cedar shingles left over from the roof of the new strawbale construction meeting room, but this is optional. It does look nice on the lids though and helps reduce the weight.

The full three bay design is wonderful for anyone with a large amount of compost to process, but it can just as easily be made as a double or single bay unit using the same technique.

Deconstruction

We won't bother with exact dimensions as they will depend on the size of the pallets you get and how much you adapt the design to fit the space available.

You will need 100+ planks to make the full size three bay version, which means deconstructing about 10-12 pallets depending on breakages of course. This may sound a little daunting but it won't be so bad if you break down the pallets as you acquire them rather than collecting them all and then doing it.

Traditional deconstruction requires the use of a wrecking bar, claw hammer, bolster and nail pullers. Alternatively put a metal cutting blade into an electric saw and just cut through the nails.

▲ 1 Trial layout.

▲ 2 Compost bins boarded up with unclad lids.

Materials List

- ▷ About 10-12 pallets, to provide 100+ planks
- ▷ 50 x 50mm wooden battens (6 per bay)
- ▷ Wooden shingles (optional)
- ▷ Nails

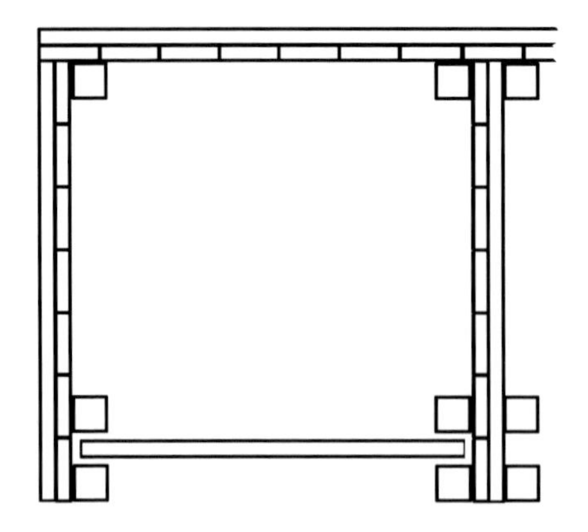

▲ Construction diagram. You can of course put the vertical planking on the outside side if you wish.

Getting Started

The back is formed by joining two large frames which are made from the tops of pallets with the base and spacer parts removed. You then need four smaller ones for the bay frames plus two frames for the lid. Lay them out as in pictures above and work out how much the bay frames need to be shortened to make the lids fit properly (1).

Putting It Together

The bay frames are attached to the back frame using internal upright battens of about 50 x 50mm carcassing which is also used to create the channels for the removable front slats. If you don't find such suitable material, fence posts would be ideal and you could even set some of them into the ground if you wish.

Nail together as required, i.e. don't nail it all up if you need to move it to a final location, just make the back and webs as separate items. You will now have a completed frame with lots of gaps

▲ 3 Adding shingles to the lids.

▲ End bay clearly showing the boarding layout.

to fill. Nail planks vertically to fill up the spaces (2). In Mark's pictures you can see he has nailed them all on the inside because he knew he was going to shingle the outsides. If you aren't going to do this you might want to plank the outsides instead for a neater finish.

The fronts are made by cutting planks to length and dropping them into the grooves between the uprights at the front of the webs. Do not fix these as you will want to remove them to empty the bins!

The lid frames do not need any additional planking as the shingle cladding will cover them completely. The shingles keep the weight down and stop the lids being too heavy to lift. If you are making the lids out of pallet planks rather than a clad frame make a lid for each individual bay otherwise they will be too heavy. In either case make sure you have a safe method of securing them in the raised position and take care when raising and lowering them. The lid frames are attached with strap hinges to the back frames.

The ends, lid frames and pillars are then clad in cedar shingles (3). You will need two bales, about 76 shingles in all. Work from the bottom up overlapping evenly and cross lapping the rows, trim off the top shingles and you're done.

Costs

Mark Fisher's total costs were just £16.27. Pallets free, buckets of nails recovered from skips and the shingles were left over from the roof of the strawbale meeting room which was being constructed on site at the time.

Footnote

We have been inspired by this design and are currently collecting pallets to make a single bay version outside the office plus a full size one off site. Shingle clad they certainly look better than most composters and should be quite durable (I have a back gate made of pallet timber that's now 20 years old). We hope you will be suitably inspired too.

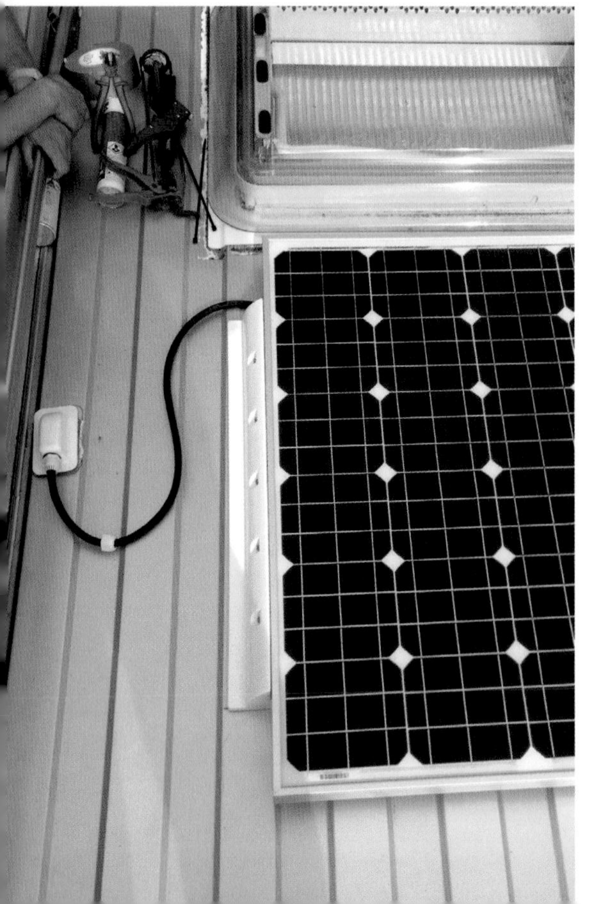

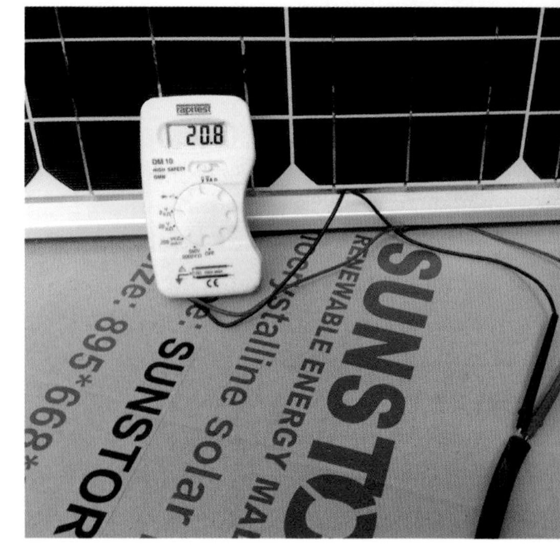

▲ (top): Checking PV panel position prior to fitting.

▲ (above): Checking the output voltage of the panel (around 21 volts is the normal open circuit voltage for a healthy panel).

◄ Trial fitting of the solar panel and cable entry gland.

7

Installing Small Scale Solar

Fitting solar panels to your campervan or other off-grid structure needn't be daunting. Here John Adams guides you through the planning stage and presents a case study of a typical installation

In this article I will describe the steps needed to set up a small scale solar PV system for off-grid living. Both examples described here are fitted to recreational vehicles (RVs, motorhomes, campervans) but the same principles apply for caravans boats, sheds, cabins, yurts, etc. The only real difference is the way the solar panels are mounted. On a vehicle or boat they will probably be glued down (though not necessarily), while on a static structure they are usually screwed or bolted down using metal brackets and a mounting frame of some kind.

System Planning

To get started we need to determine the size of the installation that is required. This is subject to a lot of variables. The main two, certainly for a live-in situation, are:

1. As a rule of thumb, how much power do you need to live comfortably for two days without recharging and without the batteries dropping below 12V?

2. How much roof space do you have where you could mount the solar panels without them being shaded, and how much sun do you expect the solar panels to get?

If you can't achieve (1) you will either need to cut your consumption (LED bulbs make a huge difference), fit a larger leisure (deep discharge) battery or fit a second one and wire together in parallel (1). I ended up doing both: changing all my lighting to LED and fitting a second battery as well. Just remember when siting batteries that they should be in a

▲ 1 Strapped down, additional leisure battery.

▲ 2 Twin 75W panels on glued down mounts.

well ventilated space and very securely fixed down.

Point (2) though will probably determine what can and what can't be done. Road vehicles and boats usually have quite limited amounts of roof space, some of which will already be taken up by openings, vents, aerials, etc. So the question is, do you have room for the area of solar panels needed to meet your requirements? Bearing in mind that the panels will never achieve anything like their rated output, as: the orientation to the sun is poor (except at midday); existing fittings will almost certainly cast shadows at some time; it isn't sunny all day every day; and in the winter things are even worse with days without sun and a low sun angle when you do see it.

Solar Panels

If you have the space, fit as much solar as you can. I had room for two 75W panels either side of my rooflight and could potentially fit another one behind it in the future. If you don't have enough roof space, the only options are to mount the panels on a roof rack frame above the existing roof clutter or to use portable panels on a frame, but these will need storing when you are moving and are easy to steal.

Personally, I rather approached this backwards. I wanted enough power to be able to live and work off grid for a few days at a time and to have both the leisure and starter batteries kept topped up summer and winter. I already had the two 75W solar panels which looked like they would fit on the roof where I wanted them, so I took it from there (if you are using conventional panels like this, remember the mounting blocks add quite a bit to the size, mine just fitted) (2). If space is tight, glued down flexible panels take less space, are less prone to damage and are ideal on lifting-roof campers and boat coachroofs.

This meant that in ideal conditions (midday in the South of France in June) I could in theory generate 150W. So dividing the 150W output by the 12V nominal battery voltage (*Watts divided by Volts = Amps*), you get 12.5A per hour,

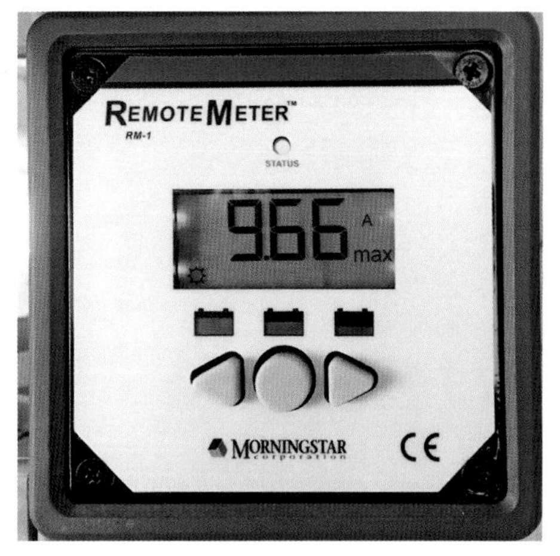

▲ 3 Morningstar Sun Saver Duo solar controller and remote meter showing a peak demand of 9.66 amps being met by the solar panels.

which decided my choice of charge regulator.

Regulator/ Charge Controller

I went for a 20A Morningstar dual output model and a monitor panel (3). As the extra cost of buying a slightly oversized controller is slight, it can be a good move, as it will then never be strained and allows for future expansion. The monitor panel was largely for test purposes and if you already have another way of monitoring your batteries you probably don't need one. What it has shown me though is that the demand from the panels is usually very low (with just one event at 9.66 amps).

Cabling

The cables from the panels to the controller are all rubber armoured twin core with 4mm (AWG11) copper conductors (this looks huge to anyone used to AC mains wiring, but DC current loses voltage very quickly so thick conductors are needed to minimise losses). If your panels output less than 100W you could get away with 2.5mm (AWG13), but I like the ruggedness of the 4mm and would use it anyway.

For the reasons above, cable runs need to be kept as short as possible and should be very carefully planned in detail before even ordering the cable. I was fairly lucky and could run my cable through the overhead lockers and then down into the base of the settee/bunk which housed the second battery and also formed the back of the bulkhead I needed to fit the solar controller on to. That was the input but the output still had to get to the leisure batteries and the starter battery which was under the bonnet. I found a route under the floor insulation and behind the step to the habitation side of the bulkhead but gave up at that point and wired it, via a fuse, into the existing in-coming feed from the battery. It is not easy to predict exactly how much cable you will need, so order at least 1m more if you are certain of your lengths and 2m+ more if you have any uncertainty.

▲ 4 Safe wiring – screwed battery connections, inline fuses and cable ties for strain relief.

Connections

A word on connectors. A modern solar panel will come with a pair of prewired MC4 connectors on short cables. If so, use this system on your extension cables. My old panels weren't prewired and I wired directly into them and used a junction box internally to join the two feeds to the single cable going to the controller (a prewired MC4 Y-branch adaptor does the same job more neatly). At other places in the system you will need to use inline blade fuse holders. These tend to come prewired so it is necessary to use a crimp connector to join them onto your cables (4). Personally I am not a fan of crimp connectors and don't really trust them; however they seem to be alright if you make sure they can never come under any strain. Finally, you need to connect the output cables from the controller to the batteries. These need to be bolted connections: never use crocodile clips. Most battery terminals come with accessory wiring points, but if yours don't, change them for ones that do.

Safety

While on the subject of batteries, do remember to disconnect the negative/ground straps from both the leisure and starter batteries, plus any AC connections you may have, before starting work. Take care not to short batteries and wear safety glasses while working with them.

Panels can be tricky to handle at heights and there is always the risk of falling off the roof, so if at all possible get a second pair of hands. Cover solar panels to stop them generating until all wiring is finished.

In Practice

Since fitting my panels I have never had any of my batteries drop below 12V summer or winter and can live and work off grid pretty much indefinitely. I hope this article and the case study help you achieve the same.

Resources

Parts for both installations were sourced from **www.sunstore.co.uk**

Solar PV Installation Case Study

Fitting a 120W PV panel and a twin battery charge controller on a Fiat Ducato Trigano Tribute 650 Campervan

Materials List

- 120w solar panel
- 10A duel battery charge controller
- Two 500mm solar panel edge mounts (these suited this installation but corner mounts are more commonly used)
- Carbond 940FC adhesive
- Cable entry gland
- 7m of 4mm rubber cable
- MC4 solar panel connectors
- Two In-line blade fuse holders and 10A fuses
- Assorted screws, crimp connectors, cable ties and insulation tape

▲ 1 Before embarking on your own project it is a good idea to look at a similar one to help plan everything. Here the *PM* team looked at the author's twin PV installation on the roof of his motorhome before starting work on fitting PV to a Trigano Tribute campervan (parked in front).

▲ 2 The existing roof bars would need removing before the 120W PV panel could be fitted.

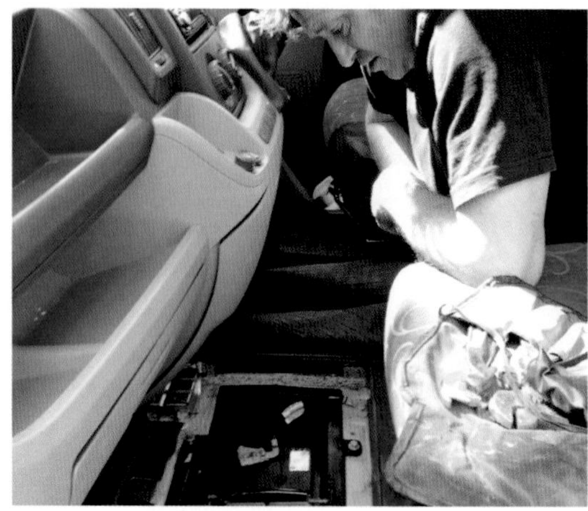

▲ 3 Finding the batteries and planning the cable routes.

▲ 4 The old roof bar mounting point allowed a ready made cable route to the external airvent on the back of the fridge. The holes would be covered by the cable entry gland.

▲ 5 Screwing the mounting blocks onto the sides of the solar panel using stainless steel screws and pre-drilled pilot holes, a job made easier by having two people.

▲ 6 Fitting positive and negative MC4 connectors on to the roof end of the cable. Colour coding helps to get it right and self-amalgamating tape seals out the weather.

▲ 7 The panel connected and the excess cabling tidied up, taped and cable-tied to the back of the panel. The cable was then fed into the van via the cable gland.

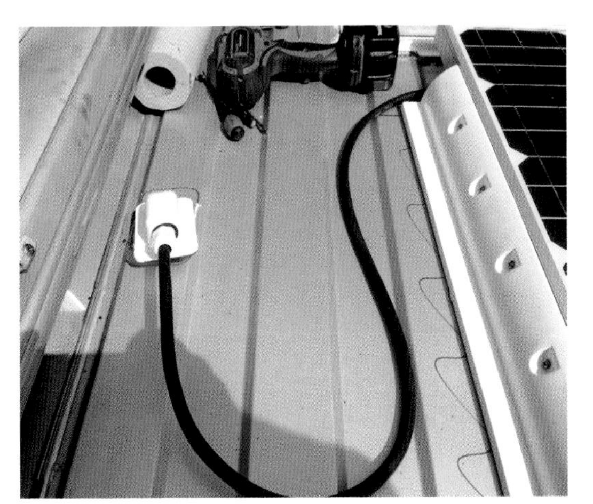

▲ 8 The position of the panel mounts and cable gland finalised and marked on the roof ready to be cleaned up for glueing but the cable laying needed completing first.

▲ 9 The removal of interior panelling was reduced to a minimum by feeding the supply cable along the existing inner roof channel and into the door frame.

▲ **10** Hole drilled from the leisure battery locker to the door frame and the cable fed in.

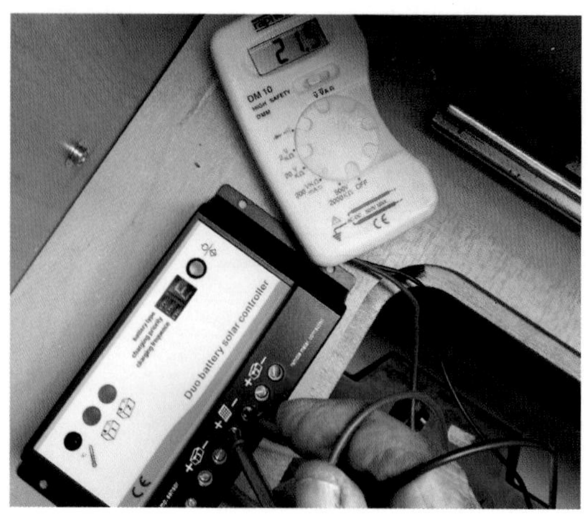

▲ **11** Cable wired into controller and tested (an input of about 21V is good).

▲ **12** The marked roof areas were solvent cleaned and abraded and the mounts and cable gland coated with Carbond 940FC adhesive (you can also use Sikaflex 252).

▲ **13** With the panel and cable entry glued down we could carry on inside, though care was needed not to move the supply cable or vehicle for 24 hours.

▲ **14** The solar panel was covered to stop it generating while the batteries were being connected to the charge controller. Note the 10A fuse fitted in the positive wire.

▲ **15** The cable run to the starter battery under the cab floor mainly went under the matting and was able to enter the battery box alongside the existing earth strap.

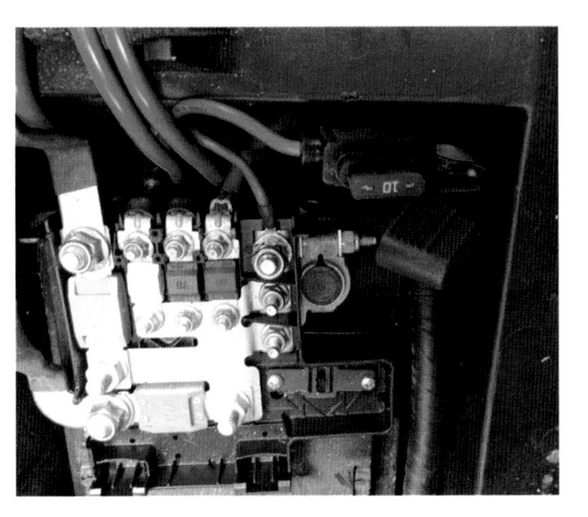

▲ **16** The final positive connection bolted down to a spare starter battery terminal.

▲ **17** Charge controller showing it all successfully installed and working.

THE NORTH FACE

8

Rocket Stove Hot Tub

Alicia Taylor and Jamie Ash explain how to build a wood fired hot tub with the added efficiency of a rocket stove

Materials List

- An oil drum – used oil drums can easily be found online
- A tall narrow central heating hot water tank – our local scrap metal merchant had lots of these
- Hot tub shell – we found one for £40 on eBay; an old bath could also be used
- Copper piping
- Pump (optional)
- Slabs or a flat surface – we recycled some slabs from the garden
- Wood for building a frame

A wood fired hot tub is a fun addition to any community or permaculture project. We wanted one that ran efficiently and was relatively easy and cheap to build. This design is what we came up with.

A rocket stove is a simple stove design with many benefits. A good air draught into the fire allows the stove to burn fuel very efficiently and at high temperatures. The fuel burns completely, which means minimum waste and a smokeless burn. The draught created by hot air rising up the chimney transfers heat directly to the top of the stove. In this case, it is then transferred to the water in the pipe, with much lower heat loss than with an open fire.

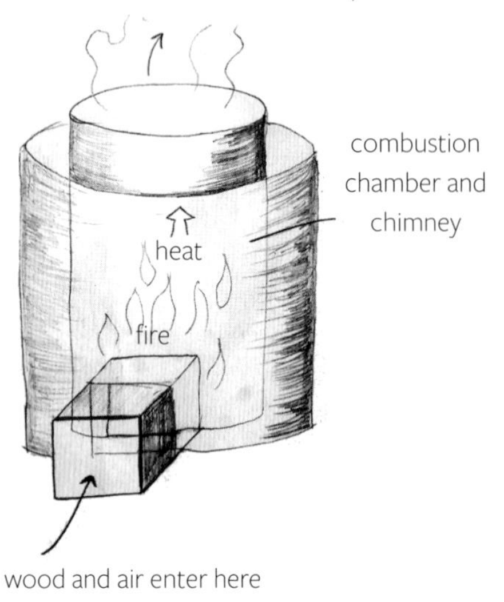

▲ The basic rocket stove design.

▲ 1 Clearing a space for the hot tub.

Firstly, clear or lay a hard, level surface to build the hot tub on (1). This stage is very important or it will be difficult to build the frame on later!

To make the rocket stove, cut the top of the oil barrel off (you will need an angle grinder for this).† Then remove the insulation from the hot water tank. This is to prevent it burning when the rocket stove is lit. Place the hot water tank inside the oil drum so the internal pipe coil is at the top (2). Carefully cut a hole in the top of the hot water tank, so as not to cut into the coil of the pipe inside.

The next step is to make the entrance to the rocket stove. This is where the fuel is fed into the stove; it also allows for good air flow. There are different options for making this. Plenty of information is available on rocket stove designs and ratios if you wish to make the stove as efficient as possible. We made a fairly simple stove but it is still efficient enough to heat the hot tub from cold in half an hour using only twigs.

We cut a 20cm square hole at the bottom of both the oil drum and hot water tank, then folded a sheet of metal to form a square tube of the same dimensions. We then inserted the tube through to the hot water tank (3). Because rocket stoves are so efficient they can run on thin twigs and sticks rather than logs, and these can be fed into the entrance tube as they burn without needing to be chopped up. Try lighting your stove at this stage to make sure it works.

▲ 2 An oil drum and a central hot water tank can easily be found online or from a local scrap metal merchant. To make the rocket stove, the hot water tank is placed inside the oil drum.

▲ 3 A sheet of metal folded into a square tube forms the stove entrance. Square holes of the same dimensions are cut in both the oil barrel and hot water tank.

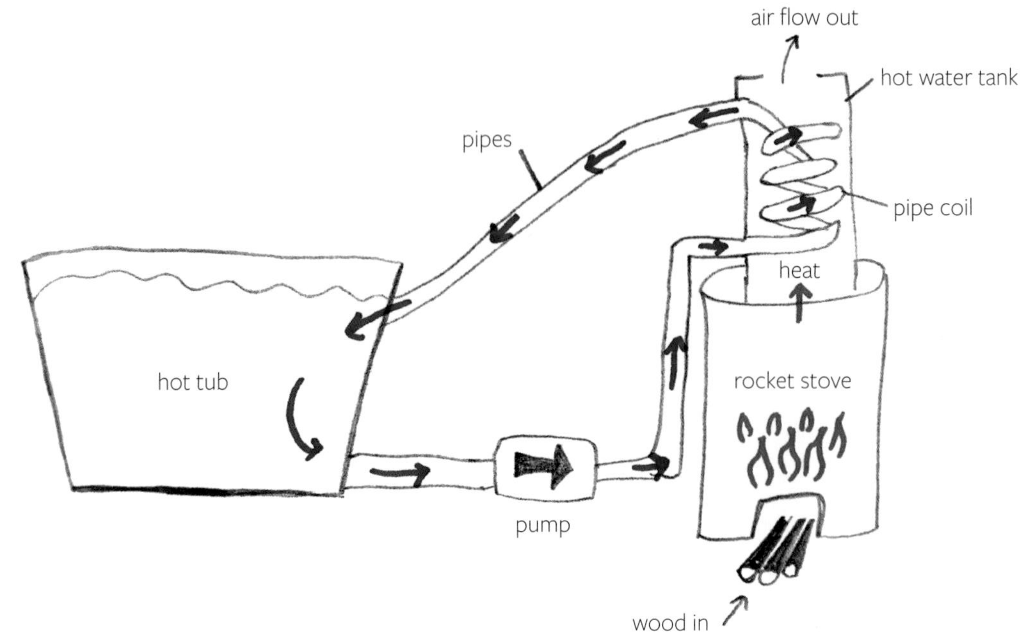

▲ 5 Diagram showing transfer of heat through system.

▲ 4 Gravel is used to insulate the stove.

Next the hot tub shell and stove are put into their final positions, next to each other, on a hard level surface. The space between the hot water tank and oil drum is filled with insulation (4). We used gravel, but vermiculite or any non-flammable insulating material could also be used.

The stove then needs to be connected to the hot tub as shown in the diagram (5). Cold water flows from the bottom of the hot tub, through the pipes and into the stove where it is heated. As the water heats it begins to rise through the piping and flows back to the tub. This creates a current of water circulating from the tub through the stove and back to the tub. The water is heated a little more each time it circulates. We added an electric

▲ Rob and Duncan test the hot tub!

pump to pump water to the stove and speed up the process. It would also work without the pump but the water would take longer to heat.

To improve efficiency, insulation was added to the piping and a lid was made for the hot tub. We used foam pipe lagging for the pipes. A lid can be made from scrap wood with insulating material on top, depending on what you have to hand. Paint it black to absorb more heat from the sun! Leave the lid on while the hot tub is heating up.

The next stage is to build a wooden frame for the tub. We made some supports to hold the weight of the water and people, then finished with cladding around the outside.

Further Improvements

A water butt could be incorporated into this design to collect rain water for use in the tub. Using a black water tank and coils of black piping would allow the water to be partly solar heated. Heat coming out the top of the stove could be used to cook or make hot drinks. This is something we are hoping to experiment with; creating a way of positioning a pan of water above the stove without blocking the air flow. When the hot tub water is dirty we use it to water the plants so the tub can be integrated into the garden system!

† Caution: make sure the metal drum is washed out and well ventilated before cutting, as a sealed drum may contain explosive liquids or vapour.

9

Geodesic Growdome

Simon Mitchell, aka SimontheScribe, shares some of his experience of building his experimental, low-cost geodesic growdomes

Ever since learning about Buckminster Fuller at art college I have been fascinated by icosahedral structures. Over a couple of years I have experimented with some of these useful structures to make growing domes for food.

Dome One

The first type I built, and the most basic shape was an icosahedron, a 20 faced sphere recognised even by the Ancient Greeks (1) – it is one of the Platonic Solids (Tetrahedron, Cube, Octahedron, Dodecahedron, Icosahedron). An icosahedron is a regular polyhedron with 20 identical equilateral triangular faces, 30 edges and 12 vertices. It is a form found in nature; for example some viruses have icosahedral shells.

◄ Dome 3 in the process of being covered.

Made from thirty hazel sticks gathered from the woods over a winter, and a length of polytunnel liner, the first growdome cost about £30. What I learnt from this one was that growing spaces need a good airflow and that if you are going to strim around them, they need to be raised from the ground, otherwise the strimmer will cut the lining. Also, one very important point: if you don't anchor them down they will blow away. This one lifted from the ground and flew for 6m, turning a complete somersault and landing the right way up, scattering all my seedlings over the ground. The dome was completely unharmed which gives you some idea as to the resilience of this structure.

Dome Two

My second geodesic growdome was a three frequency icosahedron, where each face is broken down to nine smaller

▲ 1 Dome One was part of an icosahedron made from hazel sticks.

▲ 2 Dome Two was a three frequency icosahedron.

triangles (2). I used bamboo rods and connected them with bits of old hosepipe nailed together into six-pointed star shapes. This had a good ventilation system, in fact so good that a blackbird family soon learnt how to get in and steal my strawberries, until I netted them.

The dome lasted for a couple of seasons and was made from recycled plastic sheets, bubble wrap and cling film for less than £20, which was mainly the cost of the bamboo. The joints for this dome were too flexible and it gradually turned into an egg shape under its own weight. I shored the roof up for a season with a 2m high 'strawberry tower', made of strawberry pots of diminishing size, stacked on each other.

The problem with this structure was that it was just too complicated – such a small structure didn't need so many rods and connectors. The great thing about this dome though, was that it was built on some old concrete garage walls as the base, which worked really well as a passive solar heat sink.

Dome Three

So, onto my third dome. This is the top half of a two-frequency icosahedron. I chose to go for a much stronger and permanent structure and ordered tanalised 50 x 25mm specially for the purpose, along with sand and cement to put in a 'crazy paving' base from my

Materials List

Base:

- Sand
- Cement
- Paving slabs

Dome:

- 50 x 25mm tanalised wooden struts
- Discs cut from galvanised metal sheet
- Screws
- Polytunnel cover material
- Polytunnel tape

▲ 3 Top third of Dome Three raised to make construction easier.

▲ 4 Dome Three has its legs cemented into the ground to prevent it blowing away.

garden of rocks (on the edge of Bodmin Moor in Cornwall).

I can't over-emphasise the importance of having a model to work to for dome building that has more than one strut length, otherwise it is easy to get lost in the structure. Marking the different struts helps too; I painted mine with some left-over wood treatment. I connected the struts using metal discs, pre-cut with metal cutting shears and drilled using a cardboard template. The discs were made from a sheet of galvanized metal I found, and a sheet of corrugated galvanized roofing that seemed to have been flattened by a tractor. The struts were just screwed onto the discs.

Connecting the struts and raising the dome up on tyres allowed me to connect new pieces to the underside (3). I had to give special attention to the joints at the top because they took additional strain during the construction. I bolted on plywood disks for support.

Piece by piece, it took shape and it became clear to me that a half-sphere dome would be easily tall enough rather than the five-eighths dome I had intended to build. This was fine as it meant there was wood left over for an inside structure.

I dug holes under each of the strut nexus points at the base and put in some uprights to clear the plastic lining sections from the ground and allow me to build in ventilation. I cemented these uprights in, making horizontal corrections with a spirit level as I went round (4). I used a mixture of techniques to fill in these undergaps: engineering bricks, concrete and dog food tins, old slates, carpet, bricks and blocks with cement (one side was next to the gas output of our septic tank and I certainly didn't want methane build-ups in the dome).

Then the floor went in, like a mini patio made from all the flattish rocks and bits of slate I could get my hands on. I painted some of the rocks at the back black, and left them raised to absorb more heat from the sun.

Finally, it was time to put on the

▲ 5 Dome Three nears completion. The coloured rods indicate the two different lengths required to make the shape.

polytunnel cover. I ordered another £30 worth because I intended to recycle the cover from the first dome into this one. This glazing part was quite laborious as I had to cut the sheeting into rough triangles, staple it onto the struts and then trim to size (5). It is best to work from the bottom up, then the rain will flow down, and not into the dome, due to the overlaps. Once I got to the top of the dome it got tricky and there was lots of stretching up a ladder. I had made one top triangle as a detachable window which helped with the top glazing and is essential for airflow. I also made a detachable window on one side for wheelbarrow access. All the overlaps were then sealed using a waterproof tape made for polytunnels. I made a door by taking out a cross strut and putting in hazel sticks which pushed the space open so as not to weaken the structure. This has a porch from which I hang netting to keep out the ever-present blackbirds. The first season I had a good crop of blueberries, picking salad, coriander that absolutely loved the warmth in there, squashes, tomatoes, basil and peppers. Next growing season I will hopefully have time to concentrate more on the soil quality in my pots. All in all the wood, screws, plastic sheet, sand and cement cost me around £150, a great investment for a growdome which I hope should last five years at least.

To look at these DIY growing spaces in more detail please see: **www.makeagreenhouse.co.uk**

▲▼ Dome Three offers $15m^3$ of growing space and was surprisingly cheap to build.

10

Durable Pig Ark

Stuart Anderson describes how he used sheets of recycled plastic to build a durable pig ark

We built our first pig ark out of a couple of sheets of old building site plywood hoarding bought from the Brighton and Hove Wood Recycling Project. It stood up well to pigs not only sleeping in it but also vigorously scratching themselves on the outside. The bottom edges of the ply, however, got wet, went soggy and rotted. After a couple of repairs, it was clear that we needed to renew it completely before the next lot of weaners arrived.

The ideal material would need to be able to stand up to whatever the weather could throw at it, cope with being thoroughly cleaned, stand up to the physical rigours of being home and scratching post to two or three pigs and also create a pleasant environment to bed down in. In line with our permaculture principles, it would also have to be made from an environmentally friendly material.

Plastic might not be the first thing that comes to mind when you search for green building material. It's versatile, ubiquitous and can be easily moulded but it has high embodied energy and it'd be difficult to justify its use to make something as large as a pig ark, ideal though it might be.

◄ It went down well with the new arrivals.

Plastic is, however, used for all manner of other things and so we create an awful lot of plastic waste: five million tonnes of plastic each year in the UK alone, of which an estimated 29% is currently being recovered or recycled. We might appease our consciences by popping our plastic into the recycling bin, but do we think much about what happens to it when the council lorry takes it away?

I've had the opportunity to go behind the gates at a landfill site. I encountered a dystopian scene, like something out of the film Mad Max, with gigantic bulldozers rolling on huge studded steel wheels spreading and crushing the fetid mass of detritus, with the apparent aim of squashing as much stuff in before that landfill is declared full and sealed up. Great... if you're a seagull or a rat. So we can throw our plastic in a big hole in the ground, somewhere out of sight or we can recycle it... but into what and how?

▲ Assembling the pig ark. The recycled plastic panels are cut to size and screwed to a simple wooden frame.

▲ The completed pig ark is both strong and easy to clean.

Materials List

- 2 of 2.4 x 1.2m panels of recycled plastic
- Wood for making frame
- Wide-headed screws
- Corrugated steel sheeting

Products containing plastics are, in most cases, made from multiple plastics which are almost impossible to recycle through conventional methods because of their different chemical makeup. Whilst some methods of recycling have been established in the UK such as bottle-to-bottle recycling, the vast majority of mixed plastics still end up in landfill sites up around the country and, in some cases, even overseas.

Protomax Plastics Ltd, of Somerset, have developed a process which eradicates the need for polymer separation and converts low-grade, mixed, waste-plastics into a powder which is then blended, heated and pressed into 2.4 x 1.2m x 19mm thick panels. These are marketed as an alternative to plywood sheeting (often using imported wood) and virgin plastic panels, and are fully recyclable at the end of their useful life.

Pig Ark Design

Our design for a pig ark uses exactly two 2.4 x 1.2m panels. One panel is cut down the middle lengthways to provide the two sides and the other panel is divided width ways to provide two ends. We were lucky to have a qualified carpenter, Max, volunteering at our smallholding who was at the time doing his Permaculture Design Course with Brighton Permaculture Trust.

The plastic is easily worked with ordinary woodworking tools. Although it's sold as an alternative to plywood, it's probably more descriptive to compare it with MDF in that it's made of fibres compressed together. As it doesn't have alternate layers of material with a grain at right angles to each other, it's less rigid than ply and is brittle in thin strips. We screwed the panels onto a wooden frame

▲ The inside of the ark showing the wooden frame which supports the siding and corrugated roof.

▲ As happy as pigs in ... well, plastic.

(made of very durable and environmentally sustainable black locust), which is how I had made the original plywood ark. With boisterous pigs in mind, we used wide headed screws to spread the load but you could achieve the same effect by putting a washer under the screw head. With the smaller bird and bee boxes, it was OK to screw straight into the endgrain but we found that a smaller-than-the screw pilot hole useful to prevent break-out; one has to aim accurately for the centre of the 19mm panel for the same reason and countersinking helps the screw head bury itself.

The only leftovers were the bits off the top of the arches and the cut-out to create the door. Determined not to create yet more plastic waste, we tried to think of smaller items that could be constructed out of this material and our next volunteers, Anne and Fiaz, knocked up a blue tit nestbox, leaving us most of what we needed to make a five-frame beehive.

In fact, I shouldn't have worried as broken panels and scraps can go back in the shredder and be remanufactured into new panels, thus closing the loop.

Long-lasting and weather resistant, I can think of many uses where these recycled plastic panels would be ideal, especially anywhere wet or humid, such as for sheds and other garden structures, animal and poultry housing, dog kennels, green walls, compost toilets, worktops, planters, lining trailers, etc.

Three years on since writing this article, the pig ark is still in use, stands outside all year and has required no maintenance other than cleaning with a pressure washer.

Resources

Storm Board LLP is the board manufacturing relative of Protomax Plastics Ltd.
www.stormboard.net

They supply Travis Perkins as distributors through their national chain of stores. Current price is £79 per board.
www.travisperkins.co.uk/ecostormboard

Innova Solutions sell an adhesive/sealer suitable for use with these plastic boards, called NovaSeal UltraBond.
www.innovasolutions.co.uk

Black & Decker
Workmate

11

Cider Press and Scratter

You don't have to own an orchard to make apple juice or cider – there are apples aplenty if you ask around your neighbourhood. Peter Willis explains how to make an apple scratter and a press . . . on a budget

The bumper apple crop of 2011 provided an opportunity to realise my long-held dream of making cider. I learned valuable lessons that year, and in 2012 I made 25 litres of cider and 15 litres of apple juice from 80 kilos of apples. The equipment is relatively straightforward to make, but it takes time to assemble the necessary parts on a budget, so it's worth starting early. It's also worth asking around if (like me) you don't own an orchard – a surprising number of people will have surplus apples that they will be happy to let you have in return for a few bottles of juice or cider.

In this article I will concentrate on making the equipment. For user-friendly instructions on making cider, I recommend the Wittenham Hill Cider Portal: **www.cider.org.uk**

◀ Portable apple press made from fence posts and a hydraulic bottle jack.

Author's update

I've now made cider (and one year some perry too) for 4 years with this set up, and it's still going strong. The largest volume that I made was 105 litres of cider in 2013 when there was a very good harvest and I had access to an orchard that has since then unfortunately been cut down. It took two weekends to make that quantity. What takes the time is always the pressing rather than the scratting, which emphasizes the benefit of making the press rather larger than mine. The finished product is variable – some years it has definitely been sharper than others, but it's always satisfying to sit down with a glass of home-made cider in May or June and look forward to the next harvest.

Materials List

Scratter:

- 200m, 150mm diameter, wooden drum
- Bearings
- Shaft
- 4 x 40mm stainless steel countersunk screws
- Pallet wood or wooden planks for frame, plus screws etc. for fixing
- Thin stainless steel sheet
- Small countersunk stainless steel screws (to attach the steel sheet to the hopper)
- Small pan head stainless steel screws (to locate the steel sheet in the bottom part of the frame)
- Electric motor or electric drill
- Residual Current Device (RCD)

Press:

- 100 x 100mm wooden fence posts or multiple smaller timbers for frame
- Wooden plank worktop for base plate, plus wooden strip for edging
- Plastic water pipe
- Silicone sealant
- 5m of 1.5m wide plain voile or other net curtain material
- 2 tonne hydraulic bottle jack

- Large sterilised container
- Muslin
- Sterilised bottles and demijohns

▲ 1 Screw depth gauge.

▲ 2 Drum in hopper showing staggered screw pattern.

▲ 3 Frame, drum and bearings.

▲ 4 Hopper (upside down) showing stainless steel sheets.

To make juice, you will need a 'scratter' and a press. A competent DIYer/scrounger can make them for considerably less than the cost of purchase, or for a cost comparable to that of a few weekends' hire.

Making A Scratter

Numerous websites provide instructions for, or demonstrations of, scratter projects. Most scratters involve a revolving hardwood drum studded with screws. In exchange for the promise of a couple of bottles of cider, my brother-in-law turned a beechwood drum for me. I settled on a diameter of 150mm and length of 200mm, although the dimensions don't seem to be critical. A local engineering company provided a length of surplus 25mm stainless steel tube, although in the end my brother-in-law happened to have both a pair of bearings and the matching shaft, which he fitted to the drum for me. Bearings (look for 'pillow block bearings') are readily available for a few pounds on eBay. If you don't have access to a lathe or a woodturning relative or friend, you could assemble a roughly shaped drum round a shaft, and use the bearings as the basis for something along the lines of a pole lathe.

I screwed 4 x 40mm stainless steel countersunk screws into the drum, leaving the heads protruding by 5mm (1). I arranged the screws in eight rows around the circumference, staggered so that each point was swept twice per rotation (2). The pattern seems to be unimportant, although the consensus seems to be to aim for a screw-head every 5mm along the length of the drum. I used a total of 72 screws.

The drum sits in a wooden box, open top and bottom (3). A hopper sits on top of the box. I attached it with quick-release catches, but you could use bolts with wing-nuts or some other similar arrangement. Clearances are important: I allowed 5mm at each end of the drum, and 2.5mm between the screw heads and the front and back of the box (i.e. a total of 15mm on top of the diameter of the drum). I hand turned the drum several times before fixing, to ensure that there was no contact. I fixed thin stainless steel sheet, cut from an old dishwasher, to the bottom few centimetres of the inside front and back faces of the hopper, extending down over the upper few centimetres of the drum box (4). The steel sheet is probably unnecessary, but looks professional! Slots cut in the bottom edges of the sheet fit over stainless steel pan head screws, screwed with 1mm clearance under their heads, into the front and back of the drum box, a few centimetres below the drum centre line. This arrangement keeps the steel sheet firmly against the inside of the box so that it does not catch the rotating screws.

I made the box and hopper from a beech plank left over from some kitchen worktops, but marine or exterior plywood, or a laminate work surface or table tops, would do just as well. Pallet wood would do for the frame. My frame sits on a pair of chairs, but you could make a custom built support.

Add An Electric Motor

Most of the scratters for which I found plans or videos on the internet were powered by an electric motor, typically of 1hp, via a belt to reduce the speed to the

▲ 5 DIY apple scratter (chopper/masher) powered by an electric drill.

few hundred RPM that seem to be ideal in this case. I searched for a long time for a suitable motor. Although they are readily available new, they are expensive, and I couldn't find a second-hand one at all. A redundant electric lawnmower motor might also do the trick. I contemplated using some sort of hand crank and gearing, or a treadle arrangement like a pole lathe, perhaps with a flywheel and freewheel mechanism, but then thought of using my electric drill (5). Its 520W (0.7hp) power rating is not far off that of the 1hp motors generally used, and it has the great advantage of variable speed, which I set to a low setting, and the ability to connect directly to a shaft, via the chuck. I suspect that it strains the motor a little, and I would consider buying a £40 1,000W drill specifically for the task if I were intending to make much larger quantities of juice. A Residual Current Device (RCD) is of course an essential precaution.

The Cider Press

The press was very straightforward. I made a rectangular frame from 100 x 100mm fence posts. You could use multiple smaller timbers, with whatever fastenings you can scrounge, but you should probably aim for a frame at least as solid as mine. Welded or bolted steel members would also be an option. Constrained by the available timbers, the internal measurements of my frame were 600mm wide x 540mm high. I would have preferred a slightly larger frame.

▲ 6 The scratter in action – keep fingers out!

Three lengths of 100 x 50mm, with cutouts to fit over the bottom frame member, support the base plate. I explored using the inside of a dishwasher door as the base plate, and an old washing machine drum as a press cage (I would then have made up net bags for the pulp, and a circular follower to fit the cage, along the lines of commercially available presses), but my limited metal-cutting abilities and tools made me go for wood.

I made the base plate from more of the beech plank, with plywood backing, and wooden strip screwed around the edges to retain the juice. A work surface or table top would also be ideal. A piece of blue plastic water pipe fits into a hole drilled in the edging strip, and leads the juice into a bucket. I sealed the base with food grade silicone sealant, although you may not need it if your joints are tighter than mine. You will need a few pieces of thick plywood, or other board, a little smaller than your base plate, and a former made of wooden strip a few centimetres deep, the same size as your plywood pieces. Finally, you will need some net curtain material. I bought 5m of plain voile, 1.5m wide, which I cut into 1m lengths. To power the press, having rather gratifyingly bent the scissor jack from my car when testing the press last year, I bought a 2 tonne hydraulic bottle jack, costing around £15. A more powerful jack will increase the yield a little, but will of course also increase the strain on the frame.

▲ 7 Apple pulp in a press cloth.

▲ 8 Building a stack of cheeses in the press.

▲ 9 Pressing the stack with a bottle jack.

The Fun Bit

The pressing process is very straight-forward. First, wash your apples (a garden hose and a borrowed plastic laundry basket seem to do the trick, or a plastic garden tub) and cut out any badly rotten bits. You don't need to worry too much about bruises, wormholes, etc. Feed the apples into the hopper. Place a container under the box (another garden tub, suitably scrubbed and sterilised, is ideal) and, with your drill set to a low speed, switch on. You may want a cover for the hopper – as the level drops, the apples bounce, and spray and apple chunks tend to fly (6). It goes without saying that you should keep fingers, children and animals well clear of the scratter – the idea is to produce apple pulp rather than mince!

Place the press base plate in the frame, and the former on the base plate. Spread a cloth over the former and heap pulp onto the cloth (7). Spread it out evenly, then fold the cloth over, slide the former off and place it on top of your first 'cheese'. Repeat the process until you have a tower of cheeses (8). Place a couple of plywood pieces on top of the tower, sit the jack on top and carefully apply pressure (9). You may need to slacken the jack off a few times and insert pieces of timber packing as the stack compresses. A couple of metal plates will prevent the jack from sinking into the wood. Juice will start to run into your container. Don't be alarmed if it is the colour of strong tea – it seems to lighten when fermented or pasteurised.

▲ 10 Home pasteurizing in a preserving pan.

▲ 11 Fermenting cider in demijohns.

I found that once I had squeezed out as much juice as I could, it was a good idea to fold each cheese in half and press the stack a second time.

Don't forget to wash the scratter (first removing the motor) and press thoroughly. A washing up brush gets even the most tenacious bits of apple pulp off. Then air dry wooden parts carefully – I have found that residues of juice on wood tend to grow interesting furry cultures.

Pasteurising

The squeezed pulp can go on your compost heap, or will delight your pigs. If you're planning to drink some of the juice rather than turn it into cider, and want it to keep for more than a few days, you can either freeze it or pasteurise it.

I pasteurised all of my juice in order to kill any bugs, having collected most of my apples from the ground in a public space (with permission) this year. Pasteurisation is surprisingly simple and effective. I used empty wine bottles kindly donated by a local restaurant. I washed and sterilised them in the oven, filled them to nearly full and placed them in a water bath so that the temperature in the centre bottle reached 75°C, maintained that temperature for 30 minutes, then corked them (10). It seems to keep for several months.

As for the cider, it works out at about 7% alcohol, is flat and bone dry, but very appley (11). Well worth the effort!

12

Simple Wood Fired Oven

Self-reliant suburban permaculturist, Chris Southall, shares his outdoor cooking design and build

Getting Started

I always start by accumulating and preparing my materials. If you are using recycled bricks you will need to knock off the old mortar with your hammer and bolster.

Pick the site for your oven carefully. You will be preparing food there, so being close to the house if possible is best. You will also be making smoke so I suggest not too near to the neighbours (also site it at least 2m away from flammable objects such as fences or sheds).

You can build the oven at ground level, but I find it much easier to have it raised up on a plinth about 600mm high. You can build a plinth from (recycled!) concrete blocks or similar load bearing materials. The base needs to be about 1m in diameter.

◀ The finished wood fired oven complete with metal door and cowled chimney. The cooking tools shown are chapatti pans with the original wooden handles replaced by lengths of metal tubing.

Making The Template

I built my oven to have the inside shape of a catenary curve. This curve makes an arch that doesn't need buttressing or reinforcing. To make this shape, stand your cardboard up and hang a chain with the ends of the chain the width of the inside of the oven apart, and the bottom of the loop the height of the inside of the oven down (1). In my case the width was 550mm and the height 380mm. Get your friend to draw round the chain and then cut on the line to make the former. This will guide you as you lay the courses of bricks making your oven's dome.

Materials List

- About 200 ordinary clay house bricks (recycled from a skip is fine)
- Sand and cement for the mortar, mixed three parts sand to one of cement. A little builders' lime is good as well, though not essential. If you do have some, the mix is six parts sand, one part lime, one part cement.
- A piece of cardboard, such as part of a large box, and the dimensions of the inside of the oven to make a template. Our oven is 550mm diameter x 380mm high inside.
- A chain or heavy piece of rope to mark out the curve on the template
- Something to cut bricks with, i.e. hammer and masonry chisel or bolster or an electric disk cutter
- A bricklayer's trowel, shovel and a level
- A piece of scrap metal pipe about 100mm to 150mm in diameter and a metre long for the flue
- A short piece of bed iron (angle iron) about 500mm long to go above the door
- A piece of scrap steel sheet about 400 x 240mm for the door (from an old central heating boiler, for example)
- Two chapatti pans from an Asian supermarket or old frying pans

Building The Oven

If you are making a plinth, I suggest you build a circular wall the diameter of the oven, fill it with rubble and cast a thin layer of concrete on top to make a level base. The first part of the oven is a circular floor of bricks; try and keep the bricks close together with no mortar between them and make your floor as level as possible (2).

The width of the door into our oven was 320mm and was set by measuring the oven trays we were going to use (keeping the door as small as possible makes the oven stronger). The wall of the oven is 228mm thick (the length of a brick) and is built keeping the mortar away from the hot inside face (the bricks will stand the heat but the mortar won't like it). Lay the first three courses of brick leaving a gap for the door (3).

Next, wedge the cardboard former into place and carry on laying the bricks to make the dome (lay each brick touching the former) (4). Move the former round to give you the shape of the dome as you build.

Include the angle iron door lintel above the fourth course of bricks (5). The last ring of bricks holds the flue pipe in place. Put the pipe in situ and lay the bricks tightly up to it. Finish around the pipe with a ring of mortar. A coil of wire makes useful reinforcing when buried in the mortar around the pipe.

The door is made from a piece of scrap steel sheet with an old metal handle (we used a door knob). Thick gloves are essential when handling the door or the cooking pans (the original wooden handles on the chapatti pan were replaced with long metal pipes to make them easier to handle and less combustible) (6).

▲ 1 Making the template.

▲ 2 Base layer on plinth.

▲ 3 Building up the sides.

▲ 4 Using template to help form dome.

▲ 5 Angle iron door lintel.

▲ 6 Stove pipe and door.

▲ The author cooking pizzas.

Cooking With The Stove

We burn our stove for about one and a half to two hours to heat up the bricks. The embers are then scraped to the sides and we can start to cook. We cook a series of recipes in each session. We start with something that needs a very hot oven – pizzas are ideal – they cook in as little as two minutes when the stove is at full heat. We start the dough at about the time we light the oven. We then use the remaining pizza dough to make bread rolls or small loaves. These cook after the pizzas in about 20 to 25 minutes. As the oven cools we cook a cake, flapjack, and baked potatoes (wrapped in foil). If the oven cools down too much we light a fire again for a short time (about 15 minutes) and go on as before.

Happy building and eating! Do give us feedback on your experiences. Our contact details and lots of other information about our house conversion are on our website at: **www.ecodiy.org**

Our Favourite Wood Fired Outdoor Oven Recipes

Pizzas

Made this way these pizzas are the best I have ever tasted. They need to have a thinnish crust because they cook really quickly. The metal pizza paddles are preheated in the oven and the pizzas are made from normal bread dough with a favourite topping – mine is mushrooms, home-grown tomato, pepper, onion, courgette and aubergines with loads of cheese ... mmm! I like to add mixed seeds to the bread dough for extra interest.

Beetroot & Chocolate Cake

Yes really, it has to be tasted to be believed. We like to make this in a tray so that it can cook more quickly. The tray we use is 22 x 18cm.

- 115g Wholemeal flour
- 1 tsp Baking powder
- 170g Soft brown sugar
- 1 tbsp Cocoa
- 200ml Vegetable oil
- 3 eggs
- 2 medium beetroots, cooked, peeled and puréed
- 55g grated chocolate or white chocolate drops for younger taste buds.

Mix flour, baking powder and cocoa together, add sugar and mix again. Beat eggs and stir in with oil. Finally add puréed beetroot and chocolate. Pour out into greased baking tray and bake in the oven at the cooler end of its cycle.

Seedy Fruit Flapjack

- 2 large tablespoons syrup
- 115g soft brown sugar
- 115g butter or margarine
- 250g rolled oats
- 50g seeds such as pumpkin, sunflower, mixed
- 75g sultanas or raisins, chopped dried apricots

Gently melt syrup, butter and sugar together in a saucepan, without boiling. Take off the heat and add remaining ingredients. Stir well and press out into a 25 x 20cm baking tray. Cook in a coolish oven.

13

Self-Watering Container Garden

Mark Ridsdill Smith explains how to make a genuinely self-watering container garden which prevents your plants drying out and can double your yields

If you're growing food in containers you'll know that one of the biggest challenges can be the watering. Most vegetables need lots of water to grow well. In hot weather some containers may even need watering twice a day. This is time consuming and, of course, impossible if you're busy elsewhere. Then there are weekends away and holidays: how do you prevent your plants from drying out?

Part of the solution – and an important innovation in container growing – are containers with a water reservoir. These are increasingly available commercially (Earthbox is a particularly well designed model) and there are many DIY designs on the web if you want to make your own. Reservoirs not only make watering easier, but also, by providing a constant water supply, reduce the stress on the plant. This leads to higher yields, particularly of fruiting crops like runner beans, tomatoes and courgettes. Manufacturers claim you can get up to double the yield. My experience suggests that these claims are probably about right (I grew over 5kg of runner beans in one container, for example).

But water reservoirs still need filling up, pretty much every day in summer: a growing tomato plant on a warm or windy day will drink a gallon of water or more.

So how do we keep the reservoirs topped up while we're away? An elegant and effective solution is to link the water reservoirs of several containers together with plumbing, and then keep them topped up from a water butt. The water flow into the reservoirs can be regulated by a ball valve in a control tank that works exactly like the cistern of a toilet. When the water level in the reservoirs drops, the ball valve opens, automatically filling the reservoirs.

Section Through A Self-Watering Planter

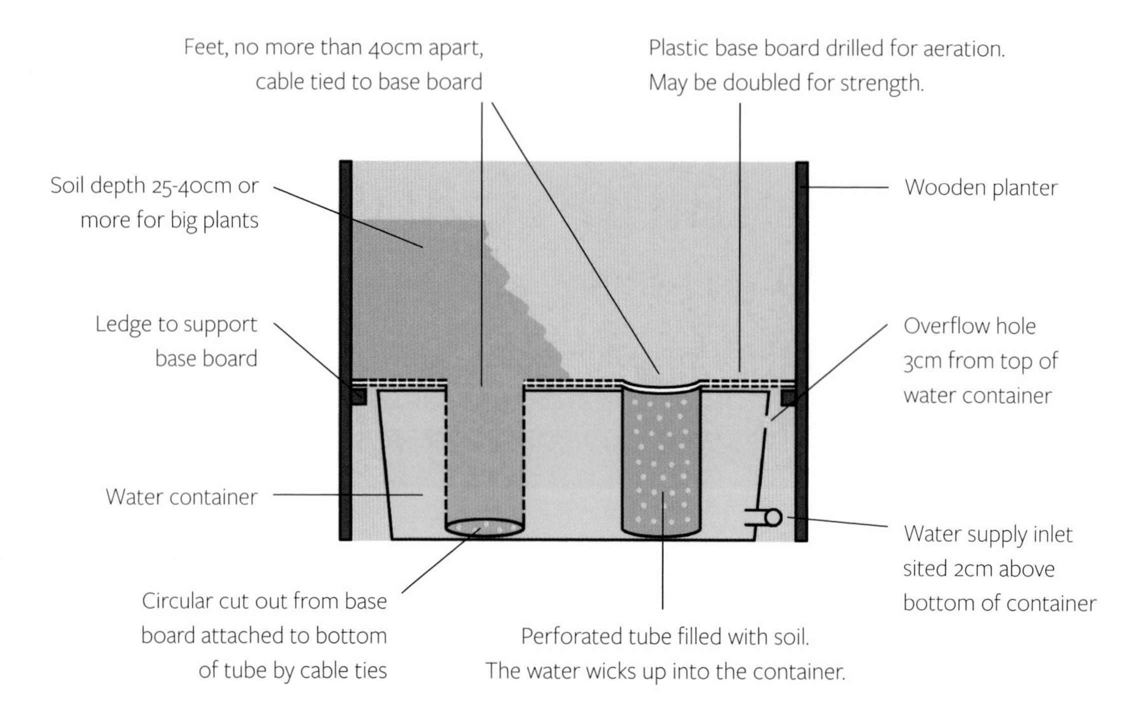

Materials List

- A horizontal surface that the containers will finally sit on (each container must be at the same level or the automatic self watering system will not work). Please also remember that containers filled with soil are extremely heavy: so you must also be certain your space can support them.
- Wood to make the boxes: I used old floor boards. You could also use plastic boxes. I chose wood in part for the aesthetics (the balcony was in public view), and in part because it let me tailor the boxes to exactly fit the space.
- Corrugated plastic sheets to make the false bottoms (old estate agent signs are ideal)
- 110mm PVC drainage pipe (often found in skips/dumpsters)
- Large plastic boxes, one for each container, to act as reservoirs. I used old 50 litre recycling boxes, plus a control tank box.
- Cable ties
- Screws
- A drill, a 110mm hole saw, a mitre saw, and screwdriver

▲ System components, (left box) planter and reservoir, (right box) planter, reservoir, feet and plumbing, (right) control cistern with ball valve.

The idea behind this system was adapted from the free, downloadable *Guide to Setting Up Your Own Edible Rooftop Garden* which is published by the wonderful Rooftop Garden Project in Montreal, Canada. Here I summarise how to build an adapted version. If you decide to build your own I recommend that you also study their guide as it goes into more detail.

You will need a few specialist tools and some practical DIY skills to build this. But anyone who's happy to tackle simple carpentry or plumbing jobs at home should have a crack. The investment of time and effort will repay itself in time saved watering and higher yields.

Construction

1. Construct the wooden boxes. Preserve the wood with Osmo oil or raw linseed oil, and line the inside with plastic.

2. Cut the top off the plastic containers so that each is 15-20cm deep. These will form your reservoirs.

3. Drill a 1cm diameter hole about 12cm above the bottom of each plastic reservoir. This is an overflow hole and prevents overwatering of your plants.

4. Drill lots of small holes in the 110mm drainage pipe and then cut it into

▲ Flow control cistern.

▲ The corrugated plastic base with the legs attached by ties.

15cm lengths (you want 20-30 holes in each length). These bits of pipe will sit, like feet, in the water of the reservoir. The water will flow through the holes and wick up into the main part of the container. The number of 'feet' you need depends on the container's surface area: they should cover 5-15% of the total surface area. More and your soil will get too wet, less and it will not be wet enough.

5. Cut some corrugated plastic card to fit snugly inside the box. This will sit on top of the reservoir and separate the soil from the water. Drill small holes in the plastic to allow air to circulate into the soil.

6. Using a hole saw, drill 110mm holes in the corrugated plastic, one hole for each foot. Keep the round 'waste' cut-outs and attach one to each 'foot' as a base.

7. Attach the 'feet' to the underneath of the corrugated plastic with three or four ring ties. Put this into the wooden box with the feet sitting in the plastic reservoir.

8. You can now add a 5-8cm diameter PVC pipe to act as a fill-tube. Simply drill a hole in the false floor and insert one end into the reservoir, leaving the other end long enough to protrude from the soil. You've now got a fully functional container with a water reservoir.

9. Or you can plumb several reservoirs together and link them up to a control tank with a ball cock. You can find the pipes and connectors you need from a plumbing store or a supplier of garden watering systems. See page 52 of the *Guide to Setting Up Your Own Edible Rooftop Garden* for full instructions at **http://archives.rooftopgardens.ca/files/howto_EN_FINAL_lowres.pdf**

The principles behind this system have significant potential for adapting creatively. If you decide to experiment, I'd love to hear from you.

▲ The self-watering containers which work on a simple demand system, controlled by a low pressure ball cock.

▲ The self-watering system achieved impressive results, with yields around twice what would normally be expected.

◄ ▲ Lovely ceramic objects made from locally sourced clay, fired in a kiln made of old newspapers, in a hole dug in the author's vegetable patch.

14

Paper Pottery Kiln

Lisa Gledhill explains some easy ways to make your own pottery, without it costing the earth

Pottery is as old as farming and they both rely on the same basic element – earth – so it puzzles me that more people don't try their hand at it. Maybe the expensive equipment puts them off or maybe they've seen too many 'Generation Game' contestants humiliate themselves trying to master the potter's wheel. But I'll let you in on a secret. You don't really need any complicated or expensive stuff to make ceramics that are beautiful, functional and, best of all, free.

All you need are a few simple modelling techniques and a kiln made from newspaper. OK, I admit you won't be able to make a full Wedgewood-style dinner service using this method but it's tremendously good fun, kids love it, and it's a great way to brighten up a dark winter evening.

First you need clay and the cheapest way is to get digging! You're looking for a layer of fine smooth earth which sticks together when squeezed. On clay soil you have an advantage but for the rest of us the best places to look are river banks and estuaries.

Sometimes you find it in unexpected places – a friend once collected some excellent clay which she spotted in a trench being dug for roadworks.

If your clay contains lots of roots, stones and debris, water it down to a runny slurry and pass it through a sieve. Then leave the sieved slurry to dry, stirring it occasionally until it has the texture of plasticine. Of course if you don't want all the fuss you can buy your clay – you'll find lots of suppliers on the internet. Make sure you choose coarse clay (the technical term is 'heavily grogged') designed for low temperature firing.

Now it's time to get creative! There's no end to the things you can make without a potter's wheel but it's best to stick to small compact items at first. Beads, pendants and wind chimes are ideal Christmas tree decorations or drawer-scenters. Children usually love modelling animals but if you want something more practical, what about candle holders, pot-feet or, simplest of all, baking beans? You could also try making small bowls or plant pots built up from coils of clay. In fact you can let your imagination run riot as long as you follow a few basic rules.

Basic Tips

Make sure your creations aren't too thick, about 1.5cm at most. Aim for an even thickness throughout. For example, a model horse could have solid legs but the body would need to be hollowed out. If you do have hollow spaces, make sure there's a hole for the air to escape when it's fired, otherwise the piece will explode. Any stuck-on bits such as arms and legs should be thoroughly blended into the main body otherwise they will fall off as they dry.

When you're handling clay, cover any cuts and don't let young children put it in their mouths as it can harbour bacteria. Finally, remember to clean up all the mess while the clay is still moist as clay dust can damage your lungs.

The next stage is to let your work dry out completely. Don't put it in an oven or near the fire as rapid or uneven drying causes cracks. An airing cupboard is OK but a normal warm living room will be just fine. Leave it at least a couple of weeks as it's important the pieces are thoroughly dry.

How to Make the Kiln

In the meantime you can start building your kiln. You'll probably need around 60 newspapers (fewer if you use broadsheets). If you use buses or trains regularly you can do everyone a favour by collecting the discarded free papers.

Materials List

- Clay
- Approx. 60 newspapers
- String or masking tape
- Ceramic or metal drainpipe
- Two bricks
- Metal grill

Take two double pages at a time and twist them diagonally from corner to corner to make a paper 'stick' (1). Now tightly roll the stick into a coil like a Danish pastry and tuck in the outer end to keep the coil rolled (2). The success of your kiln will depend on how tightly you can pack your paper. When you have enough coils, start placing them on edge, side by side, to form rings (3). The aim is to make a thick base and lid built up from concentric rings of paper coils, with a space in the centre to hold your clay. Use string, masking tape or strips of papier-mâché to hold your rings together but keep the base and the lid sections separate. It may help to wrap each layer of coils in extra sheets of newspaper for added stability as you go along (4).

The space inside the kiln should be just big enough to hold your clay objects stacked together (5). A single-skinned beehive should be enough for small objects but larger ones will need a double thickness of paper coils.

When everything is ready you can prepare your kiln site. Choose a spot away from anything that might catch fire. A vegetable patch being rested over the winter is ideal because when you've finished you can dig all the ash into the soil.

▲ 1 Collect about 60 newspapers. Take two sheets at a time and roll into sticks.

▲ 2 Coil the paper sticks into Danish pastry shapes and tuck in the ends.

▲ 3 Place the coils on edge to form tightly packed rings.

▲ 4 Build up the rings and use extra sheets of newspaper to bind them together.

▲ 5 Leave a space in the top rings for your clay objects to go in.

▲ 6 Dig a hole slightly larger than the kiln and place bricks followed by a metal grill in the bottom.

You need to make a pit just slightly bigger than your kiln with a flue to draw air into the bottom (6). The easiest way is to dig a sloping trench from the base of the pit to the surface, put in a piece of ceramic or metal drainpipe and backfill the trench (7). Now put a couple of bricks in the bottom of the pit and rest a barbecue shelf or other metal grill on top. This will let the air circulate underneath your kiln.

Stack a sizeable pile of brushwood, dry garden waste, more newspaper or any other flammable material within easy reach of your fire pit. This will be your extra fuel when the fire is at its peak.

Hold a Party!

Now you're ready to go but after all this preparation it seems a shame not to turn the firing into an event. Flames look more dramatic at dusk so why not hold an evening kiln party with plenty of hot soup and mulled wine? You can even cook over the kiln-pit once the flames have died down a bit. Children love the excitement of a firing but please remember to take the same safety precautions you would with any bonfire.

Put the base of your paper kiln on the grill in the pit and attach a paper fuse soaked in a small amount of barbecue lighter fluid. Next stack your clay objects on the base (8). The early part of the firing is the most hazardous as the sudden change in temperature will crack any weak points and can even make clay shatter. You can reduce this risk by slowly heating the work to around 100°C in a normal oven then, at the last moment, carry it in an insulated container to the fire pit. Put the lid on top and light the fire. It should blaze away spectacularly in no time (9).

When the paper starts to burn down, begin piling the extra fuel on top a little at a time. You're aiming for a long slow burn (10). You can check how things are getting on by peering down the flue. When all the fuel has burned down to an orange glow, which ideally should take a couple of hours, pile soil over the top of the fire pit and block the chimney to seal in the heat (11). Now you can finally go to bed.

Getting up in the morning to dig out the kiln pit is almost as exciting as Christmas. You never know what you're going to get! Even when you've been incredibly careful, the chances are some pieces will have cracked. Other pieces may be marked with beautiful patterns of heat-flashing (12). Through trial and error you can learn to improve the results but the surprise is always half the fun.

Brush or gently wash the ash off your work. You can add extra colour with paint and varnish. This kind of firing doesn't produce enough heat to make the clay waterproof but you can always experiment with ancient techniques for waterproofing pots such as rubbing with beeswax or pitch.

If you like the results, you might be inspired to experiment with building a simple wood-fired brick kiln. This will give you higher temperatures and more reliable results, there are plenty of books and websites explaining how to take things further. Even if you just stick with newspaper kilns for an occasional bit of fun, you can be proud that you've rediscovered one of mankind's oldest skills – the ability to make your mark using earth and fire.

▲ 7 Bury a vent pipe from under the grate to the surface and put in the paper kiln.

▲ 8 Plant markers ready for firing.

▲ 9 Put a final ring on as a lid and set on fire via a doped fuse to the bottom.

▲ 10 As the paper burns away add more fuel in the form of brush-wood, etc..

▲ 11 When it has all burnt down, bury with soil and leave overnight to cool.

▲ 12 Dig out the kiln to see how your ceramics have turned out and having removed them and the bricks etc., backfill the hole.

15

'Bespoke' Wind Turbine

Beth Tilston and Will Harley offer a step by step guide to building a wind turbine from a bicycle wheel

The idea of building a wind turbine from a bicycle wheel started, like most good ideas, in the pub. Fortunately the idea survived into the cold light of day and with Will providing innovation and mechanical know-how and Beth providing enthusiasm, a willing pair of hands and a shed to mount it on, a trickle-charge wind turbine made from a Sturmey Archer dynohub wheel was born.

We built our turbine for use at the allotment where we co-work. Being trickle-charge, the turbine will never be a constant source of power (we couldn't run the house lights from it) but it is perfect for the allotment as we aren't there all the time. Being in a windy spot on the side of a hill (aren't they always), the turbine slowly charges the battery so that on the days we are at the plot, we can listen to the radio, charge mobile phones, run LED lights and even run laptops.

◀ Finished wind turbine.

The Guide

Here's our step by step guide to building a trickle charge turbine from a dynohub bicycle wheel.

1. Find A Dynohub Bicycle Wheel

As volunteers at Cranks, a not-for-profit bike workshop in Kemptown in Brighton, we had easy access to dynohub wheels but even without a bike workshop to hand, sourcing a dynohub should not be too difficult. You can find one on eBay, at the local scrapyard and also second hand bike workshops. Modern dynohubs are available but are a bit more expensive so it's cheaper to find an old one if you can.

▲ 1 Sturmey Archer dynohub wheel with first fan blade attached.

▲ 2 Turbine showing cable led through centre of headset to avoid fouling.

Materials List

Equipment

- Angle grinder or hacksaw
- Stanley Knife
- Saw(s)
- Drill
- Paint brushes

Components

- Dynohub wheel
- Old bike with forks long enough to accept the wheel
- Two-core electrical cable
- Crimp ends
- Estate agent board or similar
- Cable ties
- Bolts
- Brake cable
- Paint suitable for outdoor use
- 50 x 50mm wooden pole to use as a mast, 2-3m long
- Large jubilee clips

Electronic Components Needed For Rectifier

- Capacitor 1,000 micro farad VH50E
- Bridge rectifier 60V 2A AQ98G
- Blocking diode N91CA 3A schottky barrier
- Veroboard on which to build the circuit

Also Required

- Step up circuit to run at 12V
- 12V battery

▲ 3 Drilling attachment holes in fan blades.

▲ 4 Fitting fan blades.

Sturmey Archer dynohubs are easily available as a great many were built, but they have fallen out of favour due to their weight and age. They are useful because a) they tend to come ready installed into a wheel into which blades can be laced (1) and b) they are a permanent magnet generator with a moving magnet and static coils. Also, being designed for bicycles they are pretty weatherproof and have bearings designed to take the weight of a person, making them very strong. They are also designed to give a good output at low rpm.

2. Find A Headset & Forks, Connect Power Cable

To hold the turbine wheel and allow it to rotate into the wind, the forks and headset were used from an old bicycle. We used an angle grinder for this task for speed but a hacksaw with a fresh blade could be used. It can be advantageous to leave the stubs of the top tube and the down tube attached to help with mounting the headset on the mast.

The only critical dimension is that the fork should be able to take the size of wheel used. A two-core electrical cable was connected to the two terminals of the dynohub, led down through the centre of the headset and secured with cable ties on the way down the fork. This allows the turbine to rotate through 360 degrees without fouling the cable (2).

3. Add Blades

We made the blade from estate agent board and cut it to be the right size to fit between the hub and the rim of the wheel that we were using (3). Any durable material with a bit of flexibility could be used – for example other types of plastic sheet or drainpipe. We fitted the blades by cable, tying them to the spokes through holes drilled in them (4).

▲ 5 Fin attached. Fin mounting arm bolted through brake point.

The way the wheel is laced will probably give you a natural angle to fit the blades. It's important to fix them all angled the same way! If you can introduce some twist to the blades, all the better (to make the tips of the blades closer to the angle of the rim and the roots of the blades closer to the angle of the axle).

4. Add A Fin

You need a fin on your turbine to steer it into the wind and keep it there. We made our fin from 9mm plywood. It's hard to give a specific size for the fin, but as a rule of thumb, make it ½ of the area encompassed by the wheel. The longer the arm it is attached to, the smaller the fin can be.

If you find your turbine does not turn into wind, increase the size of the fin or lengthen the arm it is on. You can design your fin to be any shape you want, as long as there is enough area in it to steer the turbine into the wind. We used a steel tube as the arm to mount our fin and bolted it to the steel tube in two places. The other end of the tube was drilled and bolted through the brake-mounting hole on the forks with a single bolt (5). The rear end of the fin was tied to a suspending brake cable attached to the mudguard lugs on the forks.

5. Paint Your Turbine

Ideally the paintwork should take into account the fact that the turbine is going to be out in all weathers day and night for a long period.

6. Find A Windy Location

The turbine needs to be in clear air as much as possible so apply permaculture principles and spend some time observing where the best wind is. The prevailing wind is generally south westerly over the British Isles so if in doubt put it

▲ 6 Turbine secured to mast with jubilee clips.

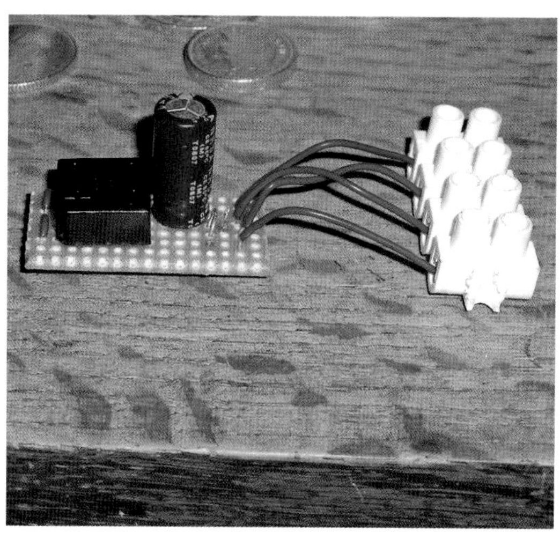

▲ 7 Rectifier.

somewhere with clear air to the south west. We used 50mm² pine as a mast to mount our turbine but you will need to size your mast to suit the weight and size of turbine and the conditions it will experience. If in doubt, go for additional strength as an over-strong mast will not affect performance. Mounting methods vary and are a matter of improvisation. We used jubilee clips around the headset and the mounting mast, tightened up to quite a high level (6).

7. Connect Your Turbine To A Battery

You'll need a battery to store all that free electricity! The most commonly used battery for small-scale generation appears to be 12V. Most cars run on 12V so there are many devices out there, which are compatible with that voltage. When you connect the dynohub to the battery, you need to bear in mind that dynohubs generally generate AC but the battery will require DC. Some older hubs have a rectifier built in but if yours doesn't you'll need to use a bridge rectifier to convert the AC to DC (7). Subsequent to this, the voltage will need to be stepped up to 12V from the 6V dynohub output.

There are instructions on the web on how to do this (see Resources). They are a bit more tricky to put together than the rectifier. The charging dynamics of the turbine are unlikely to be optimal for most batteries so battery life/performance will vary. A charge controller is a must to prevent the battery from over charging or being run down below an optimal voltage (both injurious to battery condition). Solar panel charge controllers are widely available but are not suitable for wind turbines due to the way in which any excess power is dumped. There are a variety of resources on the web concerning batteries and wind turbine compatible

▲ Allotment turbine on tall mast to ensure clear air.

charge controllers (see Resources) so do your research and follow the guidance carefully. Be careful to ventilate the battery storage area well, as charging lead/acid batteries release hydrogen, which can form an explosive mixture with air if confined.

8. Put Your Feet Up & Let the Wind Do The Work...

Resources

www.reuk.co.uk/Sturmey-Archer-Dynohub.htm

www.reuk.co.uk/12-Volt-Deep-Cycle-Batteries-for-Solar.htm

www.reuk.co.uk/Bridge-Rectifier.htm

www.reuk.co.uk/Wind-Turbine-Charge-Controller.htm

www.electronics-lab.com/project/lm2585-12v-to-24v-1a-step-up-switching-regulator

www.dimensionengineering.com

www.cranks.org.uk

16

Weave a Woollen Underblanket

John Adams explains how a simple rag rug technique can be adapted to make really wonderful underblankets and woollen rugs

I first encountered peg loomed woollen rugs when I slept on one in Ben Law's old caravan. It made for a surprisingly warm and comfortable night's sleep. I was struck by its simplicity, decorative qualities and obvious durability and wished I had one of my own. I asked Ben how they were made and he showed me a simple peg loom in his workshop and outlined the technique. The information wasn't quite full enough for me to feel confident I could make one myself and so I forgot about it. A few months later I met a woman demonstrating rag rug weaving using a small peg loom and realised this was the same basic technique. Having been taught how to make a rag rug I took the idea home and scaled it up by making a 1.5m sized peg loom for use with sheep's wool.

◄ Woven wool underbanket on a king size bed.

In this case I made the loom the width required plus a few inches either end, but shorter looms can be used if the strips are joined together afterwards.

To make the peg loom drill 15mm diameter holes to an equal depth at 32mm centres along your chosen beam. I used planed timber but roundwood would be just as good. Cut as many pegs as you have holes, these need to be about 150mm long and a good enough fit not to fall out

Materials List

- 1.5m planed timber or roundwood beam
- 47 pine dowel pegs (14mm diameter, 150mm in length)
- Drill
- Natural Jute cord
- Wool

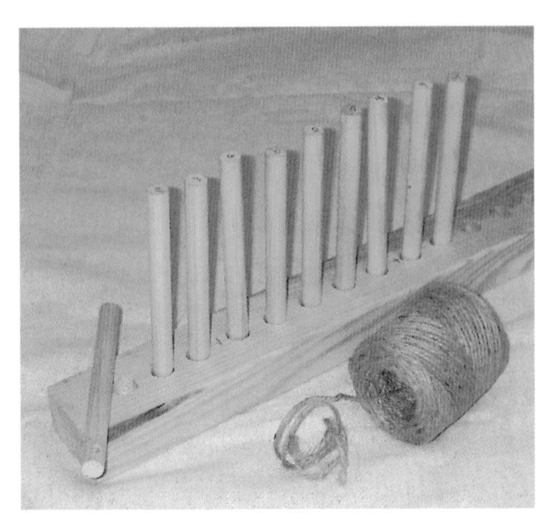

▲ 1 Pegs, beam and cordage.

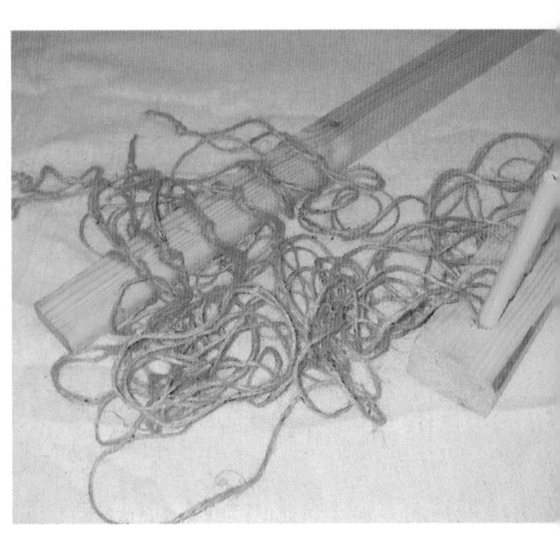

▲ 2 Cords tied off on a batten.

of the holes, but not so tight a fit that they can't be pulled out fairly easily. I used 14mm pine dowel for this job. Each peg has a threading hole drilled in it close to one end, this obviously needs to be big enough to accommodate the thickest cord you envisage using. I prefer fairly thick cordage as it is less likely to cut through the wool (1). This is particularly important if you intend to use the finished item on a floor. By now you will have 47 pegs and keeping track of them when in use can become frustrating. To save time I numbered all the holes and all the pegs. I strongly recommend you do the same as it may well save your sanity when they inevitably come adrift in the middle of the job.

The cords used were natural jute which seems to be durable while being reasonably soft and pliant. It is also fairly easy to come by – I bought mine from B&Q. You need double the length of the finished item plus about 90cm for each peg. Pass it through the peg and knot the two free ends together, then slide the peg to the opposite end of the loop. Repeat for each peg. Place the pegs in their correct holes in the beam with the cords downwards. You now have a neat peg loom ready to go and a huge mess of cordage all over the floor. To get over this I tied all the knotted ends in the correct order to a long batten (2). This allowed me to roll up all the excess and bring some order to the set up process. Later on it also proved invaluable for rolling up the woven rug as it grew. I am calling it a rug because that essentially is what was being made even though I always intended to sleep on it and made it at double bed width. Because of this excessive width I needed some way to support the weight so I clamped the beam in the jaws of a Workmate (3). This worked well and added some portability to what was becoming an unwieldy piece of equipment. If you use a short loom it can be done on your lap.

All you need now is the wool. I was lucky enough to get some Jacob sheep fleeces from my mother who has a small

▲ 3 Peg loom set up on a trestle with washed fleece in the foreground.

▲ 4 Washing machine with part of a fleece in a pillowcase.

flock in Devon but as the bottom has dropped out of the market for fleeces it shouldn't be a problem to buy some for next to nothing from your local farmer.

The advantage of Jacob fleeces is the variation of colour which produces a very attractive pattern in the final rug, though it does entail extra work as you have to be careful not to either use too much wool of one colour or to run out of a colour before the whole thing is finished.

The raw fleece will be dirty and saturated with lanolin. Most people tell you to hand wash the fleece in the bath with mild soap or soda crystals. Having tried this once and dripped the sodden fleece all over the house getting it outside to dry, I resorted to putting them in the washing machine. To do this, first put the fleece (or part of a fleece depending on the washing machines capacity) in an old pillowcase and fasten the end with a nappy pin (4). Wash on the gentlest cycle possible (wool) with something like Ecover laundry liquid. Don't let the machine spin more than the minimum required to shed excess water, otherwise you may end up with felt. Air dry outside preferably on some kind of improvised rack. You can dry them inside but they do make the house stink of wet sheep. When dry sort into colours and put back into the pillow cases or bin liners until needed.

To use the fleece look at it carefully and you will see it has natural separation lines along its length. Starting with a loose bit from one end carefully detach a strip about 5cm wide and tease it out a little trying not to break it. Take the free end and having twisted it a little weave it in between the last four pegs at one end of the beam, go round the end peg twist a bit more and weave back over what you have just done (5). Continue on to the other end of the beam, round the last peg and work back again. Try to keep the wool coming from the fleece without breaking for as long as possible. Sooner or later it will break of course and you will need it to anyway because if you have been gently twisting the wool as you

▲ 5 Starting the peg weaving.

▲ 6 Weaving on.

weaved it the fleece end will now be getting rather twisted up. To join lengths together either twist two pieces together and weave on or start a new piece by overlap weaving for a few pegs (6).

When the pegs are full, pull them out of the beam a few at a time and ease the wool off the pegs and onto the strings. Replace the pegs making sure they are in the right holes and slide the woven material down the strings. Continue until the item is complete. Then roll it up using the batten, remove the pegs and take it somewhere to be laid out (7). When it is unrolled on a flat surface untie the batten and ease the cords until the rug is square and lays flat easily. If the weave isn't tight enough it can be compacted at this stage by easing it gently but firmly from either end. If the piece is now too short it can be returned to the loom for a bit more wool to be added. Making sure you have at least 15cm or more of string at the peg end, cut the pegs free. The cord ends then need to be finished off. I knotted each pair to its neighbour and then knotted the result in groups but any other rug finishing technique will be just as good (8).

Your completed rug will now have odd ragged pieces of wool sticking out all over the place. If you are going to walk or sleep on your rug you probably won't worry about it but if you do want a neater result just throw it over a sofa and wait. Everybody that sits on it will fiddle, tucking in bits here and pulling off bits there, in no time at all it will be perfect.

My rugs have been a great success, we have one between the mattress and the bottom sheet on our bed in winter. As we have no heating upstairs my wife used to insist on having an electric blanket but I have had no complaints about cold beds since introducing the sheep's wool rug and we both sleep better. In the summer it gets used as an attractive throw on the sofa bed and we always take it to sleep on if we go camping. So for a better night's sleep, get weaving.

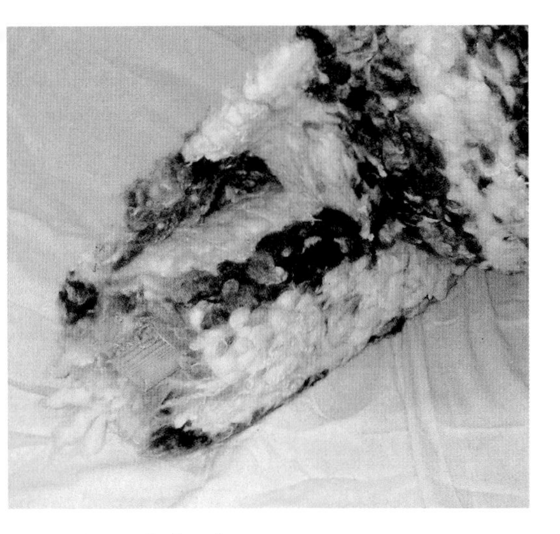

▲ 7 Rug rolled on batten.

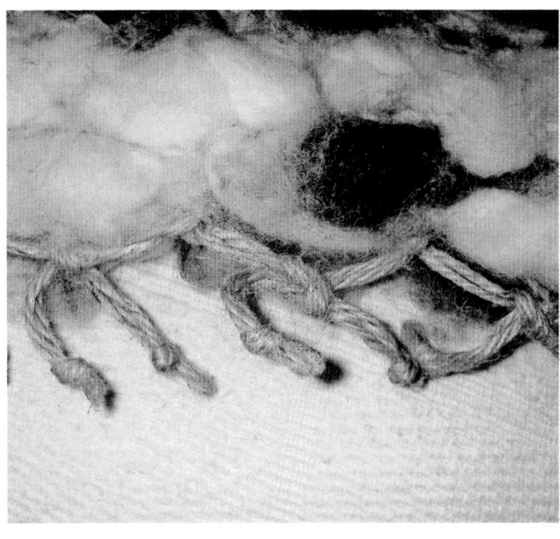

▲ 8 Knotting detail. Each pair is knotted off then tied to its neighbour.

▲ Single bed sized rug used as a throw.

◄ ▲ This very neat compost heap is used to heat the campsite shower. Last year it provided enough hot water for about five showers a day, over a fourteen week period.

17

Heating Water with Compost

Steve Hanson explains how to make enough heat for five showers a day from a pile of compost

Jean Pain, a Frenchman who has since passed away, did extensive research into what he called heat heaps. We are using his research to heat water for our outside showers used by our apprentices and students throughout their stays with us. This is about much more than taking the heat and using it productively though.

Our first try at this was done undercover in our concrete lambing pen. Constructed as a dual purpose walled pen, it is used for lambing in the spring and then as a compost bin during the summer. The area is 2 x 2 x 1m high, making the heap 4m^3, filled with organic material.

We use two ingredients in this pile. The first is oak bark chips from the local saw mill, removed from the logs before they are sawn and is a waste product (1). We buy this in by the 10m^3 load at €8 per cubic meter delivered. We use this material for mulching throughout our vegetable and forest gardens, not just for the compost heaps.

The second material is animal litter from over-wintering sheep and cows in the sheds (2). This produces a mix of organic material that is carbon rich but with a good amount of nitrogen. Nitrogen is the fire in a compost heap and the carbon is the fuel, both are important but we have found the higher the carbon level the longer the heap will stay hot. Once the heap starts to cool it's easy to add more nitrogen from urine separated at source in our dry toilets.

Building the heap is done methodically and on average takes us half a day. The first layer in the heap is 20cm of bark chips, followed by 10cm of litter. This insulates the heat exchanger from the concrete floor. The heat exchanger is 100m of 20mm rigid poly-pipe which contains about 18 litres of water. Once the foundation layer of compost is down the first coils of pipe are laid (3). Then a

Materials List

- Oak bark chips
- Animal litter
- 100mm of 20mm rigid poly-pipe

▲ 1 Bark chips ready to be mixed with animal manure and other organic matter to create the compost heap.

▲ 2 Digging out the cows litter, the second ingredient of the heat heap.

▲ 3 The first layer of waterpipe which will act as part of the heat exchanger, laid over a layer of compost.

▲ 4 Board added at front of bin and the process repeated: layer of compost, layer of pipe until full.

▲ 5 Compost temperature reading.

▲ 6 Pipes connected up to the back of the shower head. The blue pipe supplies cold mains water and the black pipe hot water from the compost bin.

10cm layer of litter is placed on top of the pipe followed by a 10cm layer of bark chips, then another layer of pipe coils (4). Once each layer of organic material is laid, the heap is wetted down so that you can squeeze water out of a handful of the mix.

The heap is topped with as much bark and litter as we can get in place without it falling over the sides. Then all the pipes are connected back up to shower fittings and the incoming mains water. The blue pipe remains delivering cold water, the black pipe delivers hot water to the shower head (5).

Once the heap has been built the microbial activity starts to create the heat we are trying to exploit. Four to five days later the heat will be higher than most people can use directly and the water will need to be mixed with cold water to make a comfortable shower.

We also build another heap to heat the water for our campsite shower. In the centre of the heaps, where the heat will be highest, we add the dry matter from our dry toilets. Yes that's right, the mixture of human faeces and toilet paper (humanure). We endeavour to monitor the heat to ensure it gets above 60°C which renders it inert (6). Once the heap has finished its decomposing process, we can use the compost.

Four years of using this system in two places on our site has taught us a great deal, both about compost making and the value of good quality compost. Like most aspects of permaculture, no one thing makes a system work and understanding the balance of what makes compost work well is the key. While many people will tell you exact amounts of carbon and nitrogen to put in your heap to make it get hot, we have learned it's not an exact science.

Bark chips alone will compost and create heat once the heap gets wet but if it gets too hot for the microbes it will kill them off. While the compost will come back to life again it will go through the same processes over and over again, taking a year or more to finish its process and become useful. This is because its tendency to overheat deters worms and other small creatures taking up residence and finishing the final compost stages.

50% bark and 50% litter will work but we have found it cools quickly and needs remaking and more bark adding to get the heap to finish quickly.

40% bark and 60% litter can get hot but not for long enough to be practical for our purposes. It seems to make a great worm farm however, and piles made to this recipe have proved to produce larger amounts of worms.

60% bark and 40% litter has proven to be the best recipe for our needs. The pile stays hot for longer with the regular addition of urine to keep the nitrogen level up. This recipe will work for 7-14 weeks, depending on the demands made on the shower. Each shower temporarily cools the heap and if people queue for several showers one after another, day after day, it will cool quickly. But it copes well with five or so showers a day for 14 weeks.

Not forgetting the icing on the cake: at the end of the composting process, we have around 3m³ of good quality, 100% natural compost from each heap. This smells fresh and clean and can be used for seed sowing, cuttings, potting up and as a general soil improver: all this on top of 500 showers per heap.

18

Collecting and Cleaning Water

Chris Southall tackles rainwater harvesting and greywater recycling at his home

People in traditional cultures place a value on being able to provide the necessities of life themselves whilst here in 'developed' countries most of us work to earn money and pay others to provide our food and shelter.

When native peoples (including our ancestors) faced problems of survival their first step was to map their territory for resources. They took a pride in honing their senses to spot game, edible plants and water signs. Here we may have lost that skill when we look for things to sustain us in an urban environment. There is no shame in asking before raiding a skip for the timber to feed your fire or build your house. Rather we can feel proud to be doing our bit to reduce land-fill, CO_2 and save the planet. That neglected apple tree on wasteland can be a source of food saving the carbon footprint of the apples in the supermarket. Once you allow your antenna to scan your environment you will, like the Aborigine, feel closer to the area you live in and spot opportunities you were blind to before. With this article I hope to draw attention to the resources that are all around us even in an urban situation.

◀ Polytunnel watered with recycled water.

Rainwater

Water is the first necessity of life; we can only live a few days without a drink but more than a month without food. Since we moved to Clacton in Essex (the driest place in England with about 500mm of average yearly rainfall) our attention has been focused on the need for water to help our plants grow. Our smallholding is an integral part of the way we live. If you don't have a garden you will end up with a different system, but the principles remain the same.

▲ 1 Rainwater harvesting.

▲ 2 1,000 litre IBC tank used to store rainwater.

Water Saving Tips

- Don't use mains water to irrigate the garden.
- Share bath water and use the left over water to wash the floor etc.
- Wash full loads in the washing machine and save washing up to do once a day.
- Don't leave the tap running when washing or cleaning teeth.
- Save your pee to use on the compost heap so you don't need to flush (probably the single most important water saving habit to adopt).

The first step on the road to abundant water is to harvest the water that falls on your property (1). Even with our low rainfall, we could store more than 125m^3 of rainwater from our house and outbuildings – more than we would ever need to use in the house – our current yearly use for four people is about 88m^3.

Water storage costs money and the cheapest storage method depends on your situation. We decided on recycled 1,000 litre IBC tanks (2). These are available cheaply from eBay or industrial estates. They do need to be protected from the sun's ultraviolet light (so we are boxing ours in with recycled timber). If you have the space, a hole in the ground is also a good storage option (3) – otherwise known as a pond! The snags are that as you use the water the water level will go down leaving you with the problem of getting the water to where you need it, and unless you have fish in

▲ 3 Water can also be stored in a pond.

your pond (which is another story) you will breed mosquitoes and become rather unpopular with your neighbours! The rainwater feeds the polytunnel by a 'leaky pipe' watering system.

Greywater

The next source is the water we pay for and use ourselves in our home – more expensive than rain but available regularly (even I have been known to have a bath occasionally!). We decided to recycle our bath, sink and washing machine water – but not our sewage.

If you store your grey water and don't use it immediately on the garden it will start to smell and if you aren't careful about the kind of soap and detergent you buy, watering can build up too many salts in the soil over a period. To avoid these problems we decided to build a small reed bed to purify the water before using it on the garden. Common reeds have the ability to provide oxygen to their roots and will use up the nitrate in the waste water as they grow. They will also maintain a colony of aerobic bacteria round their roots to help with the clean-up process. We decided on a $4m^2$ reed bed, 35cm deep, raised above ground

Materials List

- Recycled concrete
- Mud
- Old carpet
- Pond liner
- Drainage pipe
- Cold water header tank
- Hose pipe
- Gravel
- Pea gravel
- Reeds (e.g. common reeds *Phragmites australis*)

▲ 4 Reed bed wall made from rock and mud.

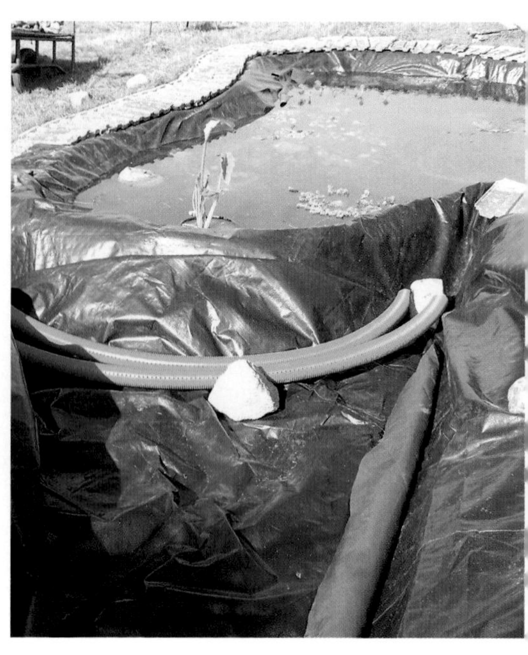

▲ 5 Lined reed bed and pipework.

level to give the height to keep the water flowing. A reed bed is basically a pond filled with gravel with reeds growing in it and a collecting pipe buried at one end to take the water out after filtering. We made ours out of recycled lumps of concrete mortared together with mud and a triangular wall cross section (4). This construction was lined with old carpet and pond liner (5). A piece of perforated drainage pipe led the water to an old cold water header tank from which we could lead it to the pond or down to a tank outside our polytunnel. The reed bed is filled with gravel starting with a layer of about 10mm of pebbles then filling to the top with pea gravel (6).

The reeds will take a year to establish themselves and their aerobic bacteria friends. In the meantime the water is cleaned by the gravel but is a little smelly due to the anaerobic bacteria in the gravel bed. I hope the reed bed water will eventually pass through the pond, but at the moment it contains too little dissolved oxygen to be good for the fish (we don't want to rely on an electric aerator). To complete the system we installed a surge tank made from a (recycled) water tank to take the water from the house. This has a crude filter made from a pan scourer (which needs to be cleaned monthly). The water passes from the tank into a hose pipe which leads to the reed bed.

I think the reed bed and pond are an attractive addition to the garden which will look even better next year as the reeds grow.

The water systems have served us well this year which has proved to be another very dry summer here in Clacton.

▲ 6 Reed bed filled with gravel.

▲ The pond and reed bed after one year.

19

Reciprocal Framed Roofs

Legendary low impact roundhouse builder, Tony Wrench, describes his latest building methods for low cost dens and dwellings

I didn't build my first low impact funky structure till I was over 40. The reason? I thought building dwellings is what experts do, and, specifically, I didn't know how to make a simple, cheap roof that looked good and kept the rain out. Walls are easy, you think, but a roof – what if it falls in?

Here is a way of making a good cheap roof that you can build as soon as you decide that the society we live in is not going to solve your housing needs for you, and get round to building your own shelter, or at least start with one for the chickens or pet dog.

Reciframes

The term 'reciprocal frame' was coined by Graham Brown of Findhorn in about 1987, so it hasn't been around for very long. I will call them reciframes to save syllables. The idea has been around for ages. Leonardo da Vinci designed a floor using reciframes where none of the floor joists were as wide as the floor itself. In Japan, there are many ingenious structures using the principle, and those Morris dancers among us will remember the Sword dance where six swords are held out into the centre of a circle by the dancers and placed one on the other until they are locked together. The flash young handsome dancer (there's usually one) then leaps up onto the sword reciframe in the centre and is supported at shoulder height by the rest to the accompaniment of wild cheers and windblown Scots bagpipe music. Yes children, try this at home.

The first reciframes I heard of in Britain were the whisky barrel houses at Findhorn community, and Jack Everett's

◀ Completed reciprocal framed roof building with cordwood walls.

Materials List

Structure

- Approx 20 x 5-6m long poles (ash, oak, chestnut, or spruce, larch and Douglas fir are most ideal)
- Nails
- Steel fixing strap

Walls

- Strawbales or cordwood
- Glass for windows

Roof

- Hazel strips
- Cleft chestnut
- White canvas
- Straw
- Rubber pond liner
- Spiral drainage pipe
- Polyfelt geotextile
- Turf

dojo near Stroud. Once the idea of a reciframe is grasped it is very easy to make your own, and the idea is spreading very fast as people learn how easy they are to make, how efficient and strong they are, and how great they look from underneath (and above, until you put the roof on). I made one as soon as I heard of the idea in 1990 and have built about 14 reciframe roofs, always using roundwood. There are about six or seven devoted reciframe builders in this part of wild west Wales, and we develop our designs by taking a good look at each other's buildings, covertly asking each other how we did a particular new feature, then amending our design a tiny bit for our next project. One day we will all meet at the same time and then the whole thing will really take off, but at the moment there is a lovely sense of spiralling evolution. I have never built one exactly the same way as the last. The whole process is improving, week by week.

Here I will go through the basics as I understand it with you, on the assumption that there will be some person somewhere who already has a much improved system. So don't treat this as gospel – use it for tips to save you time – maybe a couple of decades.

Dens & Dwellings

You can use a reciframe to roof over a structure of any shape. You don't need a central pole, and none of the rafters need be as wide as the building itself. You also get the option of a skylight in the centre, which is a big plus for a simple den or dwelling which wants to make the most of natural light. You can use from 3 rafters to 16 or more. If your building is square and you put the rafters at the corners, it is going to look almost like an ordinary roof. If your building is shaped like an amoeba, and even with walls that rise and fall as they go around, you can use a reciprocal frame for that too. The rafters don't have to be the same distance apart. What is a good idea, however, is not to try to fix the pitch of the roof. The pitch (slope) depends upon the diameter of your building, the thickness of the rafters where they meet, the number of

▲ A The advantage of putting the roof on first is that all further work can be carried out in the dry.

▲ B Overlapped henge joints being eased to fit. These are fixed to the uprights with a vertical steel pin.

rafters, and the shape of the building. But you are basically arranging rafters so that they support each other where they meet.

We hold up the rafters with a temporary support – a Y-shaped post or two lashed poles called a Charlie – near the centre while we put the rafters up. When all are in place we take away the support and the roof finds its own level. A reciframe roof is usually of lower pitch than a 'normal' roof, and you are not trying to hold a particular pitch. So most of the weight will be going straight down the walls, rather than outwards. This means reciframe roofs are good for structures such as simple strawbale dens with load-bearing walls, although I try to build in a ring of some kind to spread any outward push from a roof. I will describe my default way of working, and will try to mention options where possible.

Walls

You need something strong to put your roof on. Reciframes are ideal for turf roof construction because if you have a large sheet of rubber (preferably) or plastic it will lie, within reason, on any kind of irregular shape. Reciframes are not the natural choice for a tiled roof, for example, because of the irregular and wacky angles that you get as you approach the centre. A rubber sheet forgives all that. But a turf roof means you have a lot of weight up there, so your roof needs to be well supported. I have used load-bearing straw bale, load-bearing cordwood, and a henge of wood that bears the load with sides filled in with various materials, usually straw or cordwood. I now have come down firmly in favour of the henge system, for several reasons.

First, the weather. If you go flat out to build a nice wood henge, then put a reciframe roof on, you can have a waterproof covering over the whole building area as you start the tedious bit of building gorgeous walls of your chosen material (A). You have much more leeway if the walls don't have to hold the roof up. If you need good hard solid load-bearing walls before you get a roof on you have to mess around with big plastic sheets to keep the walls dry when it rains,

▲ C A lovely reciprocal framed roof structure with henged supports and diagonal braces.

then wait till they are strong enough to proceed, then worry that you didn't wait long enough as you pile the turfs on.

Second, the aesthetics. A wood henge looks gorgeous, specially without black plastic all over it. So I'd go for a load-bearing skeleton every time these days. My default design is round, with a level henge of good strong, plumb, posts and beams thick as your thigh in a circle, with feet buried firmly at least 70cm in the ground, charred and bitumened. The number of rafters is the same as the number of upright posts, so the weight will go straight down. The cross pieces of the henge are half lapped and pinned to give strength to the circle (B), and there are several diagonal braces around the circle to hold the skeleton solid (C).

The non henge option is, of course, possible. The walls in this option need not be level, and you can position the rafters over parts of wall that are the strongest i.e. not over windows and doors. This option requires a strong wall plate to spread the roof weight around the structure. On a straw bale wall I use, for preference, round oak branches that are curved to follow the line of the wall. If you use flat sawn wood for this, as you do on a cordwood wall, you can join the sections of wall plate better, but you need to make a saddle of hardwood for each rafter to support its weight over the middle of the wall, rather than at the outside edge. Each saddle is made from a piece of hardwood as thick as your calf and as long as from hand to elbow, cleft down the middle, and a curve cut out, with adze or chainsaw, to take the rafter. Just keep the saddles on hand and fit them as required.

▲ D First four rafters held in place by tripod and Charlie.

▲ E Manoeuvring heavy rafters into position requires great care.

Preparation & Measurement

Decide how wide your building is to be then cut your rafters (and cordwood for the walls, come to that) at least six months in advance. Ha! We are not doing the quick visit to B&Q here. We are building a funky roof of natural materials for almost nothing. Best way of obtaining roundwood to build a largish structure like a den, cabin or house is to borrow some money and buy a piece of forestry. 0.8 hectare would be more than you need. Thin the piece of forestry, cutting out lots of lovely straight poles and snedding them at leisure. Sell the forestry again, for slightly more than you paid for it. Buy the rest of your building materials (mostly pond liner) with the profit. That's the way to get a free designer home!

Anyway, I digress. However you do it, decide early and try to get your wood at source, i.e. on some forsaken boggy windswept piece of forestry in winter. Don't let me put you off – ask a friend or a specialist to get the wood for you, if you are clean out of welly boots and muscles. There is no best wood for reciframes – most wood is suitable – but bear in mind that the wood needs to carry weight across its length, so use wood with a good tensile strength that will retain those same properties after a few years. Sycamore, the poplar family and alder, for example, don't score very well in this category (although they are fine for cordwood walls), but ash, oak, chestnut, or spruce, larch and Douglas fir do.

How wide can a reciframe like this be? Well, it depends on the number and strength of your rafters and of your assembly team. A reciframe is surprisingly strong, but it is best to be safe than sorry. Don't expect more than a 3m span of a length of calf or biceps-size roundwood without a support. This assumes a manual, fairly funky way of working. The building we made in Portugal in 2007 was of eucalyptus, which is a dense hardwood. All the rafters were as thick as your thigh and probably weighed at least 100 kilos each (D) and (E). The building is 10 metres wide so we figured that we

▲ F Simulating the roof construction on the ground is a good idea.

needed three strong internal supports at the halfway mark along the three strongest rafters to help the reciframe. We also needed a team of at least eight strong young people to lift and manipulate them. For a small rural bedsit den, 5-6m diameter is a good size. I would not go for a building wider than 6m unless I used some internal uprights as well, although this is easy to do. What feels handleable for a group of people working to build a reasonable sized den is 12-13 rafters on a building of 5-6m internal diameter. Cut each rafter way too long, and cut too many as well, so you have a good choice. For my default structure of 13 rafters I usually cut about 15 rafters of 3m, the radius of the building, plus 1m over in the centre plus 1 or 2m over at the eaves. So we have 15 x 5-6m long poles, seasoned for at least a few months to reduce risk of hernias. You'll also need long poles for the Charlie and central tripod, so cut around 20 poles while you are at it.

Simulation

If you are building a den as one of a group, or if you have not built one before, it is important that everyone involved understands the principle of what is going to happen, and the things to look out for in the operation. I would therefore recommend that you always carry out a simulation on the ground before you start balancing heavy beams around above people's heads. What I tend to do these days is a two stage simulation, using the actual rafters.

1. Lay all the rafters out on the site, preferably before you have built

▲ G Measuring the height of the simulated spiral on the ground.

▲ H Working out the positions of the rafters using a short Charlie.

anything, dug any holes or even marked anything out, but after clearing the site of topsoil etc., so you have a clean flat area to work in. Bring in each rafter and lay each one across the last one about 50cm from its thinner end, making up a simple spiral on the ground (F). Reject any poles that are not sound in any way at this point. Arrange the rafters where you think you will use them, and mark on the ground where your walls could go allowing for well over 1m of spare rafter at the eaves. This allows you to see whether your ideas about how much space you have, how large you want it to be, etc. are realistic, bearing in mind what rafters you have available. Obviously you want to use the strongest and straightest parts of the poles that you have, and can even cut the thinnest pieces off your rafters if you are clear they will not be used. When you have your spiral on the ground, measure its height (G). Call it Spiral Height and entrust the sacred number to someone with a good memory, or even write it down. It will be necessary to calculate the height of Charlie.

2. Now simulate an actual reciframe with the rafters you have selected, still on the ground, but using a small Y-shaped pole or tripod Charlie maybe 1.5m high. In the picture the Charlie is a tripod (H). Since we are now about to simulate the erecting of our reciframe with real big rafters, it is necessary to explain the positioning of the Charlie, and the relation of each rafter to the next.

What we are aiming for is a central circle that is over the centre of our building. You can do what you like of course, but this is the method I know. To have a regular central hole we need each rafter to miss the centre by the same amount. We will have either a henge or a wall to guide us, and we can decide how big our hole should be. Say we want it 1m wide. Then each rafter will miss the centre of the hole by 0.5m, right? So if we put a stake

▲ | Central circle in reciframe from underneath with spiralling hazel stems covered by white canvas.

in the centre of the circle, place a rafter on the henge point where rafter no.1 will go (it matters not where we start) and place it across the circle, to the right of the central stake by 0.5m. Where does it point to across the circle? Place something brightly coloured at that point.

Our Charlie must support this pole and must not get in the way of other rafters, so the best place to locate the fork of Charlie is along the line of this first rafter, about 60cm towards the outer circle. Place Charlie in this spot, lay the first rafter on him with 1m or so to spare beyond the fork, and tie the rafter to Charlie with string. Now take rafter no.2, go around to the next point clockwise where you want a rafter (if you have a henge, it will be over the next upright), and lay no.2 onto no.1, allowing the same distance from the central stake. Put a 15cm nail into no.1 to stop no.2 rolling down no.1. Don't bang it too well – this is still only simulation – better to reuse the nails. You can tie it to no.1 as well. Note where the rafter points across the circle. Usually I find that the Sacred Red Jumper (SRJ) has to be something like two-thirds of a post span to the right of the post opposite. If you move the SRJ to the same equivalent place for each rafter as you put each rafter on, you get a regular central hole.

Card Trick

At this point we must introduce a new word. Imagine that your rafters, laid one over the next, look rather like a hand of cards. You want to see the picture or value of each card in your hand, so you spread them out evenly with the same amount of each card showing. Similarly,

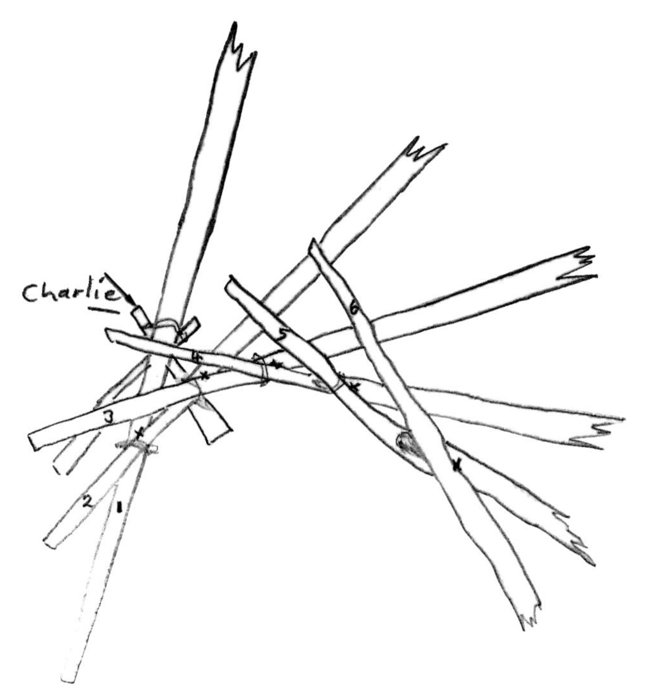

▲ Diagram showing the Charlie supporting rafter number 1 while the remaining rafters are added to the reciprocal framed roof structure. Each 'x' marks a 150mm nail. The distance x - x is a 'rummy'. Note position of Charlie just outside the circle.

we want a regular distance between one rafter and the next around the circle so that all are taking roughly the same amount of weight and each is resting on its neighbour in a regular way. We therefore use the word 'Rummy' to describe the distance between the point at which pole A rests on pole B and B rest on C, as seen directly from above. If the rummy is small, say less than a handspan, your central hole will be rather small – maybe less than 60cm across. If it is two handspans or more the central hole may be quite large – over 1m – depending on the number of rafters you use. The hole shown here (I) is of twelve rafters, with a rummy of one generous handspan. It is about 90cm wide, thus accommodating two fine coach windscreens, one on top of the other. It is regular because each time we placed a new rafter on the preceding one, we drove a long 15cm nail into the rafter below where the new rafter rests against it, and the nails are a handspan apart. (See diagram above for the rummy principle in essence).

Dear reader, I fear I have lost you. Recipes are hard work, are they not? But if you simulate your reciframe with a small Charlie as described, you will, magically, see what I mean. Using the SRJ around the perimeter, and having a rummy of the same distance between nails, you will achieve an awesomely satisfying central hole. This is kind of important, because you may well find your eye drawn to this roof on many occasions in the future. I could show you photos of satisfying and less satisfying circles, but hey, life is short. Here's a nice one, using these principles from above (J) and below (I).

▲ J Central circle from above showing additional metal strapping.

▲ K The Charlie stick is removed by hammering away at the foot.

As an alternative to the Sacred Red Jumper you can mark clearly on the top of each rafter where you want it to rest on the one below it (I put a cross 60cm from the rafter tip) and where you want the next one to rest on it (I make a groove, about 2cm deep and angled at about 27 degrees from the centre line of the rafter) whose centre is a rummy distance down from the cross. A rummy is usually about 26-28cm, to give a hole of about a metre across. More detail in my new book *A Simple Roundhouse Manual*, available on Amazon and Kindle, or from my website **www.thatroundhouse.info**

Going For Real

OK you've done your simulations. Choose a clear day and go for it. Take care lashing a good central tripod together so that you can climb up and have your head and shoulders sticking up above the circle, i.e. the top of your tripod must be high – maybe 3 or 3.5m up. Use plenty of rope. Lash your Charlie to this. If Charlie is a pole, the fork of the Y must hold the first rafter high enough for all the other rafters to pile on each other, forming the central circle, and for there to be enough room left for the last one to fit in under the first one. I use a formula: Charlie fork must be Henge (wall) height + Spiral Height + 30cm, above the ground. If instead of a Charlie pole you use two poles lashed together in a scissor-like arrangement, the point where they cross must be this height above the ground. Lash the Charlie in such a way that you can untie it easily and remove Charlie bit by bit without the feet of the support tripod getting in the way. Check this out on the pictures. Let the centre of Charlie be half your projected hole diameter to the right of the centre, and maybe 60cm towards the outer circle, along the line of the first rafter. In words this sounds like gobbledegook, but you've simulated it, so you know how it goes.

When all is ready, go for it. Have all children kept well clear. Allow no-one within the circle who is not totally focused and in full possession of their senses. Cut out chat. One person speak

▲ L Making the central circle secure with metal fixing strap.

at a time. Put rafter no.1 on, as you did in the simulation. Tie it to Charlie. Tie it loosely to the henge. Place no.2 rafter on no.1, with its heavy end clockwise around the outer wall, and its top end pointing to the Sacred Red Jumper. From your commanding yet apparently suicidal position at the top of the central tripod, whack a good strong 15cm or so nail into no.1 to stop no.2 rolling away. Tie no.2 to no.1. Move SRJ around clockwise to its next alignment place on the henge or wall. Repeat until you have put the last rafter in under the first and everything looks wonderful. With luck you will have a nicely shaped central hole and a small gap between first and last rafter. At this point, take a ratchet fixing strap and tie it around first and last. Check that all rafters are tied to their neighbours with baler twine. Untie Charlie from rafter no.1. Then get trusty and brave helpers bit by bit to hammer away the foot of Charlie or to knock gently his two feet apart (K). As the whole reciframe lowers, tighten the ratchet strap to take up the slack. When you are within a few fingers' widths of them touching, put a nail in the appropriate place in the last rafter and steer it down if you can. It makes quite a difference in what direction your helpers knock the feet of Charlie, so bear this in mind. Adjustments are much easier to make at this stage than before the whole frame settles. If you are happy with it, let it go right down. Charlie will come loose and can be removed, and you can fix the whole central circle with metal fixing straps (L) and (J). Then make good around the walls, using metal fixing strap, long nails, rebar or oak pegs. Put the kettle on.

▲ M Radials take the weight of the turf roof.

▲ N Hazel stems have a very decorative effect.

▲ O Drilling and screwing cleft chestnut at eaves.

▲ P Covering roof with rubber pond liner layer.

▲ Q Pinning the rubber membrane down behind the eaves as the turf roof nears completion.

▲ R Boarding the roof with sawmill slabs is an alternative approach.

▲ The completed reciprocal framed roof structure.

Finishing The Roof

All I have just written here takes only a couple of days to do. One for simulation and one to do it. The rest of the roof might take you two or three weeks, but here is the process in brief, and pictures. Usually we use some kind of hardwood struts, fixed as radials, to hold the weight of the turf and to bind the whole thing together (M). It depends what materials you can use sensibly. The last two roofs we have done were as follows:

1. Drill and nail hundreds of hazel strips to the rafters in a spiral (N). Drill and screw turf retainer, or fascia, of cleft chestnut at eaves (O). Cover with white canvas (I). Cover with 30cm of straw. Cover with rubber pond liner (P) and (Q). Fix spiral drainage pipe onto rubber with gaffer tape. Cover with Polyfelt geotextile. Cut hole and place windscreen skylights over turf. Weight down skylight at corners with turfs.

2. Mark out, saw, drill and screw planks, sawmill slabs etc. onto the rafters to give a solid wood ceiling (R). Fix slab fascia to eaves. Cover with agricultural fleece, then 30cm straw, then Polyfelt, then rubber pond liner then Polyfelt then turf. Add skylight to taste.

Making a simple roof like this is very cheap, it can be lots of fun to do, and they look so natural. Keep thinking along safe lines as you're working up there, but don't be afraid. Reciframe roofs have a lot of fail-safe built in, and they are much easier to make than you'd think. Good luck.

20

Making Homemade Paints

Award winning cob building pioneers Adam and Katy Weismann, share their low cost recipes for beautiful natural paints

Before it was possible to purchase manufactured paints, people made simple paints out of locally available materials, such as lime, clay, milk (casein), animal blood, urine, vegetable oils and naturally occurring pigments from rocks and the earth. These paints were healthy for people, the environment and the buildings onto which they were applied. The basic ingredients used were non-toxic, some even being edible. This means that they did not emit harmful substances, nor did they pollute the environment when they were made and disposed of. They were also highly compatible with the traditional buildings onto which they were applied, such as soft bricks, stone and cob (earth). This is because they were equally as porous as these building materials, allowing for a healthy exchange of moisture into and out of the walls, creating, in effect, a 'breathing' building.

◄ Applying alis clay based paint to a window reveal.

These simple, homemade paints are currently enjoying a resurgence. We are increasingly becoming aware of the negative effects of many modern building materials on our health and on the environment, as well as the detrimental effects that synthetic finishes can have on old buildings, because of their non-porous nature, and therefore their inability to 'breathe'. This resurgence is also being fuelled by the increase in new-builds that utilise 'natural' materials (many drawn from traditional techniques), such as earth (cob, rammed earth), straw bale, hemp-lime, and the many breathable, 'eco-friendly' building boards now available. Many of us are also becoming increasingly empowered to once again become more self-sufficient and resourceful.

We have provided two basic home-made paint recipes – limewash and alis clay paint. These are most suitable for application onto plasters and renders made out of earth and lime. A limewash can also be applied onto a cement render to provide more depth and beauty to the wall. Limewash can be applied both internally and externally. It is simply made by mixing together either a hydraulic powdered lime or lime putty with water. Other ingredients, such as linseed oil and casein powder (milk protein) can be added to make it more water resistant and durable.

Mixing Old With New

The natural, homemade paints described here are not suitable for direct application onto a gypsum plaster. Plasterboard that has been primed, however, can provide a suitable backing for an earth plaster. This can then be coated in either a limewash or a clay alis paint. A homemade primer can be created by mixing together manure, coarse sand and wheat flour paste, or it is possible to purchase a pre-made non toxic primer from most eco-building supply stores. It is not generally considered good practice to apply lime plaster onto plasterboard.

The addition of natural earth pigments to limewash will create soft and unique colours. Alis clay paints were traditionally used on adobe buildings in North and South America. At their most basic, they are made by mixing together naturally occurring clay-rich sub-soils and water. The addition of fine silica sand and wheat flour paste (a mixture of wheat flour added to boiling water), will make the finish more durable and more stable, and the addition of mica flakes and chopped fibres, such as hemp and straw will provide texture and beauty.

Wheat Flour Paste Recipe

Flour paste should be added to the mix immediately prior to it being used. Add after the clay and sand have been mixed thoroughly together. Making wheat flour paste is a simple process, requiring a cooker or small stove with enough heat to achieve and maintain a rolling boil. It should be used immediately after being made, or can be kept for up to one week in a fridge.

Method:

1. Whisk together 1 cup of white wheat flour and 2 cups cold water into a smooth paste resembling thick pancake batter. Work out all lumps.
2. Bring 6 cups of fresh water to a rolling boil in a large pan.
3. Maintaining the rolling boil, add the wheat flour and water mix slowly to the water, stirring constantly.
4. Maintain the boil and stir constantly, until the paste becomes thick and slightly translucent. The paste is now ready to use. Allow to cool down before using.

▲ First steps in making limewash: adding lime putty and water to bucket.

Basic Limewashes

A basic limewash can be made out of either a non-hydraulic lime putty or a natural hydraulic lime powder.

Basic Limewash With Putty

Use as a general limewash for indoors and outdoors.

Ingredients:

- 1 part mature lime putty (minimum 3 months, 6 months to 1 year old for best quality).
- 3 parts clean water.

Mixing:

1. Fill bucket 1/3 full with lime putty.
2. Add water gradually and whisk together using a paddle attachment onto a drill, until the consistency of full fat milk is reached.
3. Add pigment if desired and whisk thoroughly until well incorporated.
4. Can be used immediately, but will improve if left to stand for 1 hour or more. If it is standing for a day or more, more water may need to be added before use to thin it down to the desired consistency.
5. Whisk thoroughly immediately before use, and continue to stir at intervals throughout the application process.

Amounts and coverage:

6-8m^2 per litre. Coverage may vary depending on the texture and porosity of the wall surface.

▲ Power mixing the limewash.

Limewash With Hydraulic Lime (LHL)

Can be used where continual dampness is an issue.

Ingredients:

- 1 part natural hydraulic lime powder (NHL) which produces good results in most situations.
- 3 parts clean water.

Mixing:

Always wear a mask when mixing natural hydraulic lime powder.

1. Follow instructions as per mixing lime putty to reach desired consistency of full fat milk, replacing the putty with natural hydraulic lime powder.

A limewash made out of natural hydraulic lime is best used immediately.

Amounts and coverage:

7-8m^2 per litre. Coverage may vary depending on the texture and porosity of the wall surface.

When mixing and applying limewash, gloves, goggles and protective clothing must be worn, because lime is highly alkaline and therefore caustic. When working with alis clay paints, however, it is not necessary to wear gloves or goggles. In fact, the clay is positively beneficial when in contact with the skin, and therefore a joy to work with.

▲ Applying pigmented limewash.

▲ Carefully painting the first coat of coloured limewash.

Limewash Troubleshooting

1. Cracks and crazes:
   - ▷ Wall not sufficiently dampened before application.
   - ▷ Limewash made too thick.

2. Powders off the wall when dry:
   - ▷ Limewash made too thick.
   - ▷ Limewash applied too thickly.
   - ▷ Too much pigment.

Applying Limewash

Always protect your eyes and hands with safety glasses and gloves as lime is caustic. Limewash needs to be applied in a minimum of three coats (five or six is best for newly painted walls), in order to develop depth and even coverage.

1. Dampen wall prior to application with clean water. If applying onto a freshly lime plastered or rendered wall, application with a pure limewash (basic mix) can be carried out 'al fresco'. The limewash is applied when the lime plaster/render is still wet and not fully carbonated. The porosity of the limewash will enable carbonation to occur simultaneously in both the limewash and freshly applied plaster/render. This can strengthen the bond between the limewash and the wall, and can intensify the colour.

2. Always agitate the limewash in the container before application, and regularly during application. This will prevent the ingredients from settling out, and ensure that the pigment is thoroughly integrated.

3. Use a wide paint brush 100mm. For smooth walls, a short, stiff bristled brush works best. For 'knobbly' walls, such as old cob/stone walls or a harled finish, a longer bristled brush works well to access between the undulations, and to cover the larger surface area. Use a range of smaller brushes for edging and detailing.

4. Limewash is runnier than conventional paint, so care should be taken not to load the brush with too much material, otherwise it will drip excessively. By dipping the brush only half way into the limewash and squeezing out the surplus paint against the bucket edge, or flicking it a few times into the bucket, the brush will not become oversaturated.

5. Apply the limewash by brushing it vigorously into the wall surface, getting into all the nooks and crannies. Brush strokes can be in all directions, or in a tight circular movement.

6. Apply each coat thinly to prevent crazing and cracking, and avoid the temptation of going over areas already covered. The limewash will appear translucent on first application, and will only develop its opacity as it begins to carbonate (1-24 hours).

7. The first coat should be thinner than subsequent coats, which can be made progressively thicker.

8. Allow each coat to dry for at least 24 hours before applying subsequent coats. Mist each previous coat lightly before applying a new one.

Alis Clay Paints

Alis clay paints are best for interior work, and are generally applied in two coats. The recipes below are given as a guideline only, and should be refined by carrying out individual test batches on the wall. Each test patch should be allowed to dry fully before being analysed and amended according to the outcome.

It can be mixed up in buckets, a wheelbarrow, plasterer's bath or cement mixer, depending on the amount being made. Use immediately because the binding power diminishes the longer it sits, and the wheat flour paste will mould.

Guidelines for Test Patches

Testing can be done to assess the quality of the mix and for achieving the desired colouring. Make up as per recipe guideline and apply to the wall substrate (mist wall before applying). Allow it to fully dry and check it with the following in mind:

1. It should not dust off onto clothes or skin. If it does, there is too much sand, not enough clay, or you need to add more wheat flour paste.

2. The finish should be smooth (especially when polished for the final coat). If it is excessively grainy, there is too much sand or the sand is too coarse.

3. If the alis cracks excessively on drying, there is too much clay, not enough sand, it has been applied too thickly, or applied onto a dry, thirsty wall substrate. The paint should go on the wall smoothly and evenly without lumps and without dripping.

▲ First steps in making alis clay paint: make up wheat flour paste, add clay, (and sand).

1. Base Coat

Alis clay paints are very similar to earthen plasters, and exact recipes cannot be provided because of the many variations of raw materials.

Starter recipe:

- 1 part clay subsoil, coloured clay or a bagged clay (kaolin clay).
- 1 part finely ground mica or fine sand (silica).
- ½ part wheat flour paste.
- Pigment optional.

Mixing:

1. Make up wheat flour paste and allow to cool.

2. Add roughly 2-3 parts cold water to every 1 part wheat flour paste, to make it liquid enough to enable easy integration of the remaining ingredients.

3. Add sand/mica and clay to wheat flour paste and whisk together, until a creamy, spreadable consistency is achieved like thick cream.

4. Add pigment when well mixed and mix evenly.

Amounts and coverage:

6-7m² per litre. Coverage may vary depending on the texture and porosity of the wall surface.

▲ Power mixing the alis clay paint.

2. Top coat

Use the same recipe as for the base coat, except that the sand is omitted. Mica can still be used as a decorative element, and chopped straw can be added to provide texture.

Starter recipe:

- 1 part clay subsoil, coloured clay or a bagged clay (kaolin clay).
- ½ part wheat flour paste.
- Pigment optional.
- Small amounts of chopped straw or mica flakes for decoration and texture in final coat (optional).

Mixing:

1. Make up wheat flour paste and allow to cool.
2. Add roughly 2-3 parts cold water to every 1 part wheat flour paste, to make it liquid enough to enable easy integration of the remaining ingredients.
3. Add mica and clay to wheat flour paste and whisk together, until a creamy, spreadable consistency is achieved like thick cream.
4. Add pigment when well mixed and mix evenly.

Amounts and coverage:

8-10m² per litre. Coverage may vary depending on the texture and porosity of the wall. Two coats should be applied, with the second coat being polished or buffed for a smooth finish.

▲ Brush applying alis clay paint.

▲ Polishing the final coat of alis clay paint with a damp sponge.

Applying Alis Clay Paint

1. Lightly mist the walls prior to application. Apply when fresh plasters have fully dried.
2. Work the paint well into the surface of the wall, brushing in all directions. For small, curved areas such as niches, it can be mixed thicker and applied by hand.
3. Allow the first coat to thoroughly dry before application of the second coat.
4. Apply the second coat as above.
5. Allow second coat to dry until leather-hard but still damp, then use a damp, well-wrung sponge dipped into clean water (warm water will assist the process). Polish the alis clay paint with the sponge, using tight circular strokes. This should remove brush marks and smooth out the surface. It will also expose flakes of mica or chopped straw if used. Rinse the sponge often to keep a clean edge.
6. If large flakes of mica have been used, a final polish with a dry cloth will further enhance the shine. Save any leftover material for future repairs. If straw or flour paste have been added, the material should be dried into 'cookies', stored in an airtight container, and rehydrated for future repairs and fresh coats.

21

Solar Electric Bikes

Michel Daniek explains how to transform your humble steed into a solar powered electric bike

We all use our cars to drive short distances, a lot, don't we? Just to pop into town, do some shopping, see a friend, or to go to work...

I often felt guilty for my laziness in not using my bike for these little trips. It is strange because I hate being dependent on petrol and cars and my bike has always been there waiting for me in the garage. I guess my laziness has always been bigger than my guilt. A sentiment you might understand better if you had ridden on the rough tracks we have here where I live in the Spanish mountains.

Then my wife had a very good idea ... Why not convert my bike to an electric bike – an e-bike!?

I could add my 12V solar power skills to get it recharged ... a solar-e-bike.

◄ Using an e-bike conversion kit, these once standard bikes are now power assisted by an attached motor.

Suddenly a long held dream of mine became a reality, silent mobility with self-made green energy! Everyone who tries it comes back smiling; it's really addictive!

You too could get rid of that guilty feeling and be an independent, free and easy rider, and hopefully inspire others to do the same.

The first step is to change your old bike (if you have one) into a modern e-bike by using a conversion kit. Then later you can add a solar system, or a combined solar and wind system to recharge it, to make it a real solar-e-bike.

Materials List

- E-bike conversion kit (includes motor and battery)
- Wind generator and/or 2 to 4 solar panels
- Voltage charge regulator for e-bike battery

▲ 1 A typical rear hub motor.

▲ 2 A 1,000W motor allows extra loads to be carried.

Converting a Bike

Most of this technology comes from China, where it has been widely used for many years. Perhaps because they use it themselves, the quality seems to be better than what we usually get from this region. The required parts can be sourced from good bike shops, or via the internet.[1] It's possible to convert almost any bike, but it's best to use a strong, stable one, with effective brakes!

The conversion kits are all well pre-fabricated, so fitting them is usually not too complicated, and should require only a small number of tools.

The prices for ready-made kits range from €350 (£300) to €700 (£600) (depending on the size of motor and battery). And if you aren't able to do it yourself, you should budget for around €150 (£128) to €500 (£428) for professional help (depending on your bike and the mechanic).

To avoid disappointment, the size of the motor, battery, and recharging system will all need to be carefully chosen to suit your personal requirements.

Motors

There are many different motors available (1), for example:

- 250W motors will work well for sporty people who want to pedal all the time and only require up to 50% support.
- 500W motors will offer up to 75% support. These can be strong enough to work alone in flat areas
- 1,000W motors mean you don't have to pedal at all, even on steep hills!

The latter is recommended if you want to carry children or heavy loads with you (2). Be aware that these motors can power your bike up to 50km/h! In some

▲ 3 Batteries can be mounted either in the frame or on a rear carrier.

countries going this fast will require special permission, or alternatively you can get a controller with a 25km/h speed limiter fitted.

Battery

The battery can be fitted either on the rear rack or somewhere inside the frame (3). There are a wide variety of modern lithium batteries available, but the Li-Fe-PO4 are the best for e-bikes at the moment. They are available in different sizes. I recommend you use one in the range from 36V/7Ah up to 48V/20Ah.

The best size for you depends very much on the motor you require, the distance you want to ride, the height difference you need to manage, your own weight and last but not least the state of the roads you will be travelling on.

The motors can usually run with 24V, 36V or 48V but they will have different power outputs depending on the battery tension.

For example, the 1,000W motor (usually on 36V) produces 1,333W on 48V. Consider carefully, before you choose the size of the battery. You can get a range of between 5km and 80km depending on the battery size, motor, landscape, weight and road.

In my case I use a 1,000W motor with a 48V, 12Ah battery. This allows me to go up to 15km on bad roads in the mountains, even with one of my kids on the back seat. With a 250W motor and the same battery you could go as far as 60km if you pedalled with it.

Speed is usually controlled via either a twist throttle, like on a motor bike (only legal in some countries), or a load sensing switch to support you only when you pedal. The battery will stop working automatically via its BMS (Battery Management System) when it is empty, so it's a good idea to fit a battery power indicator to see how much range you have left.

▲ 4 Conceptual recharging stations.

Recharging

Lithium-Fe-PO4 batteries have no memory effect and can be recharged at any time. The battery cells are rated for up to 4,000 recharges, but realistically you can only expect to recharge about 1,500 times.

As a first step you can recharge yours from the grid with the 230V charger unit, which usually comes with the conversion kit. It makes sense to do this, so you can see how often you are using your e-bike. With this information it's then possible to calculate the right size of the solar system (if you don't have one already), in order to make a solar-e-bike out of it and make it really ecological.

Recharging with Green Energy

In the second step you can choose between a direct solar and/or wind recharging system. You will need daylight of course and/or a good wind to charge. All you need is a wind generator and/or 2 to 4 solar panels and a special voltage charge regulator for the e-bike battery.[2]

Or you may decide to have an indirect solar and/or wind charging system. This would have its own back-up battery, controller and 230V inverter to enable you to use the normal mains charger unit. With this system you could recharge your bike even at night or during periods of bad weather.

Costs

The price for a direct charge solar system, with 2 to 4 solar panels, will be around €600 (£515) to €1,200 (£1,000).

The current price for an indirect charging solar system, with its own back up battery, regulator, and 230V inverter is between €1,000 (£850) and €3,000 (£2,572).[3]

So a solar-e-bike will cost you anywhere between €350 (£300) and €3,700 (£3,172)! But let's say that with €1,750 (£1,500) you will have a very good ecological transport system with green energy! Isn't that a

▲ E-bike conversions enable powered assistance across all types of terrain.

similar price as an old second hand car?

If you use your solar-e-bike hard every day the battery will last you for 3 to 4 years (bike batteries cost about €25 (£21) to €100 (£86) a year.) So together with the lifetime of the back-up batteries in the solar system (if you choose this system) these are the only running costs you will have.

In times of Peak Oil I believe it's smart to get at least one alternative form of transportation for yourself.

If the price of petrol suddenly rises astronomically high, you'll be thankful to have a solar-e-bike to save petrol and money!

And Finally

All electric bikes are a bit sensitive to heavy rain, because of their electronics, so it's best you keep them somewhere dry when not in use. You could build yourself a small bicycle garage incorporating solar panels and/or wind-generator, recharger and back up battery (4).

My wife and I wish you great fun with your solar-e-bike and many smiles on your way.

1. Try **www.bicycle-power.com** to find suppliers in your region.
2. See Michel's blog post: **http://tinyurl.com/charging-from-solarpanels** or contact Michel at: solarmichel@hotmail.com
3. For details of how to make a solar system and for more information on solar controllers see the latest edition of *Do It Yourself 12 Volt Solar Power*, published by Permanent Publications.

Ben Law sitting in one of his handmade bentwood chairs in the outdoor kitchen. From the cover of *The Woodland Way*

22

Bentwood Chair

Ben Law describes how to construct attractive and comfortable garden chairs from small diameter coppiced roundwood

Chair making is a detailed craft that can require patience. However, most people begin with simple rustic versions such as slab chairs and stick chairs before moving on to the more complex types. One of the main differences is that the rustic stick chair is usually made from wood in the round that hasn't been quartered prior to construction. These chairs can be quick to make and some will last well. Bentwood chairs take the rustic stick frame and add curves made of woods such as willow and hazel. These chairs are attractive, easy to make and can be surprisingly comfortable.

Materials List

Frame

- 13 sticks for the stretchers: 600mm long x 38mm diameter
- 2 sticks for the front legs: 405mm long x 50mm diameter
- 2 sticks for the back legs: 760mm long x 50mm diameter
- 1 stick for the top bar: 900mm long x 50mm diameter
- 2 sticks for the diagonal braces: 650mm long x 38mm diameter

Bentwood

- 3 rods for the bow back: 3,048mm long x 16mm diameter
- 6 rods for the arms: 1,700mm long x 16mm diameter
- 13 rods for the seat/back rods: 1,500m long x 16mm diameter

Fixings

- Silicon bronze ring-shank nails: 30mm and 38mm
- Galvanized nails: 75mm

- Danish oil

Recommended tools

- Bow saw
- Cordless drill
- Small hammer
- Secateurs
- Spirit level
- Protractor
- Carpenter's square

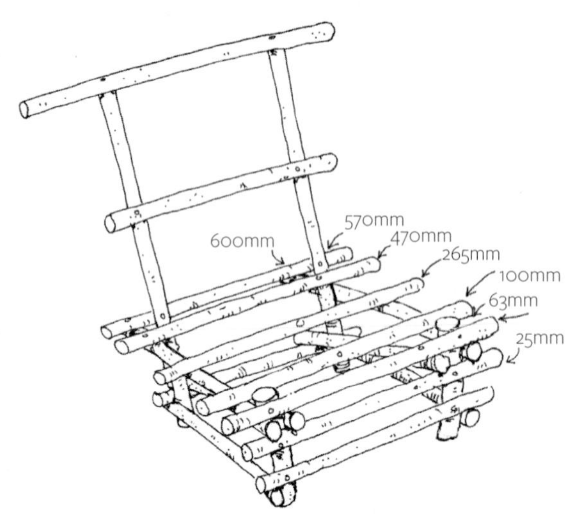

▲ 1 The frame layout showing centre measurements.

▲ 2 Laying out the side frames using a straight-edged piece of timber to level from.

Materials You Will Need

I mainly use sweet chestnut for the frame and hazel for the bentwood. In the UK, hazel and willow are the favoured species, while in the US willow is the dominant species used, with occasional chairs being made from alder or cottonwood.

Assembling The Frame

First, assemble the stick frame of the chair (1). Find a straight-edged piece of timber to use as a level line from which to lay out the side frames. The front legs should be straight and the back legs should be angled back at 12 degrees off vertical.

The top stretcher on each side of the frame should be slightly angled towards the back legs whereas the lower stretcher should be level (2). Drill the side frames and nail together with 75mm galvanized nails.

Drill and nail the cross stretchers. Check with a spirit level that the sides are level with each other (3). Drill and nail diagonal braces from the front top stretcher to the back bottom stretcher (4). Drill and nail them into the legs as well as the stretchers (two-way nailing) (5). Attach seat stretchers, a top bar across the top of the back legs and another stretcher across the inside of the back legs, midway between seat stretchers and top bar, to complete the frame (6).

Adding the Bentwood

Start with the bow back. I use silicon bronze ring-shank nails of 30mm and 38mm. These are very thin, so a 2mm drill bit (size 1) is ideal for pre-drilling. Fix one end to the inside of the lower stretcher of the side frame where it meets the back leg and bring it around in a curve so that it will attach to the back of the top bar. This first piece must attach to the top bar, leaving room outside of it to attach the other two parts of the bow back (7).

▲ 3 Using a level to check the frame.

▲ 4 Diagonal braces create stability in the frame.

▲ 5 Two-way nailing of the diagonal brace. This helps make the frame rigid.

▲ 6 The completed frame.

▲ 7 Beginning the bow back.

▲ 8 The completed bow back.

▲ 9 First rods for the arms.

▲ 10 The arms completed – note the way each rod twists over the previous one, from front to back of the chair.

Work the bentwood in your hands to even out the bow. Spend some time getting the first bow looking good, as the others will follow its shape and the look of the chair is often defined by this curve. When satisfied with the shape, fix to both sides of the top bar with 38mm silicon bronze nails and fix the end to the inside of the opposing lower side frame stretcher.

Next, fix the second bow back rod to the lower side frame stretcher against the first bow back rod. As the second bow back rod twists around and follows the profile of the first, drill and nail it to the first bow-back rod at approximately every 200mm. Repeat this process for the third bow-back rod (8).

Attach the arms to the inside of the front lower cross stretcher. Position the first arm rod approximately 150mm from the centre of the front leg and then bend it so that it attaches to the front upper cross stretcher, curves around and is fixed to the top bar inside of the bow back. Again, work the bentwood to ensure the curve is both aesthetically pleasing and comfortable for the user's arms. Match the curve of the first arm with the opposing arm on the other side of the chair (9). Getting a good match with the arms is important for the balance and aesthetics of the chair.

Attach the second arm rod to the inside of the lower cross stretcher next to the first arm rod. Twist it around the first arm rod before finishing on the bow-back side of the first arm on the top bar. Fix the second arm rod to the first arm rod at approximately 200mm intervals in the same way that you fixed the bow back rods. Fit the second arm rod to the opposing arm and then follow the same pattern with the third arm rods (10).

The last part is to infill the seat and

▲ **11** The completed chair.

back. The pattern for this infill can take many forms, which is part of the beauty of this bentwood technique. At this stage you can let your imagination dictate the pattern. I chose a pattern that brings the seat rods in at the centre of the back and then opens out again as it meets the bow back (11).

The process of bending the rods from the seat to the back should be done slowly as there is a chance the fibres could break. Start with the centre rod and then work out the spacing to keep the infill of rods even on each side. Tuck the rods under the front upper cross stretcher and then fix them to the second seat cross stretcher and to where the seat rods curve upwards. Fix the rods to the middle back stretcher and then take the top of the rods behind the bow back. Fix the top of the rods to the back of the bow back curve. Drill and fix into the back of the lower bow back rod and then cut off the surplus with secateurs, so that the seat rods do not disturb the view of the bow back when looked at from the front. Drill and fix the seat rod through the middle bow back rod as well, so they are double fixed.

Finish the chair with Danish oil. These types of chairs need to be kept inside during bad weather if they are to give you a few good years of use.

Project taken from *Woodland Craft* by Ben Law, published by GMC Publications.

23

Natural Swimming Pool

David Butler recounts the experience of creating his own

Around ten years ago, my partner Alison and I were lucky enough to buy an old derelict barn with two acres of land in Norfolk. The barn is still to be fully renovated but we have been living there in its half built state for the past four years. My energies have been diverted to a far more exciting building project: three years ago I started making our swimming pool.

Conceptual Beginnings

I have always thought that it must be possible to build a swimming pool that doesn't use chemicals to keep it clean. One summer, I had seen our water butts either choked with blanket weed or turning peagreen with other algae. Except for one: the neglected one with couch grass growing in it. Pulling up the floating mat of grass revealed stunningly clear water. I read a book on reedbed sewerage systems and realised it was basically the same biology as my couch grass water butt algae killer. Instead of reeds taking out the nutrients, it was couch grass.

◀ The plants edging the pool keep the water clean and clear, and suitable for swimming.

Surely it must also be possible to use other plants to clean a swimming pool? Searching the web to confirm the originality of my concept dashed all pretensions of genius. It had all been thought of before. Peter Petrich had been making them, along with others, for twenty years over in Austria and Germany. His company, Biotop, had made hundreds of them.

I couldn't afford to have a pool built professionally, so building it myself was the only option. At that time, in 2007, there was very little information available for self-build swimming ponds, so it was all a bit of an experiment.

Swimming Pool Zones

The Natural Swimming Pool (or Swimming Pond) is divided into two equal area zones: one zone for plants, the regeneration zone; and one zone for swimming. The plants have only sand or gravel to grow in so their only chance of getting nutrients is to take it from the water. Then hopefully the algae, like blanket weed, have little left to feed on. The regeneration zone is separated from the swimming zone by a submerged wall. This is to stop the plants colonising the whole pool.

Materials List

- Cement
- Concrete blocks
- Fleece underliner
- Plastic liner
- Fleece overliner
- Sandbags
- Flexible drainage pipe
- Shingle
- Geotextile membrane
- Sand
- 'Soft rooted' pond plants

▲ 1 Swimming area defined by block-work wall.

Planning & Digging

I decided on a swimming area of 4.5 x 11.5m, about 2.2m deep, with a shallow 3m wide regeneration zone all around it. I needed an area roughly 20 x 15m. I chose one corner of the field sheltered by a bramble filled bund. I was also able to align it north – south, forming a pleasant sun trap at the south side against the bund.

I hired a man with a digger for a few days and eventually I had a basic shape. My original intention was to build the wall from sandbags filled with sand and clay from the hole. But this was a disaster. When it rained the bags became squidgy, and started slithering and slumping until the wall gently collapsed. I tried again, this time filling them with clean sand. These were more stable but the sunlight started to turn the synthetic sandbag material into something no harder than tissue paper. They split and sand trickled out like 25 kilogram egg timers. The wall was punctured with sandbag-sized empty husks and heaps of sand.

Building the Block Wall

I reluctantly had to start again, this time digging out some foundations for a concrete block wall. I flung the sand from the sandbags into the mixer to make the concrete for the foundation. After a week I had built a block wall on the foundation, with solid 440 x 215 x 100mm concrete blocks, five courses, to just over 1m high (1). The void behind the wall was packed with sand and rammed with a tamper (a heavy metal lump on the end of a broom handle), left to settle, and rammed again over a period of weeks. This was to make sure the outward pressure of the water was not going to push the wall over.

The land was free draining, containing some clay but mainly stony and sandy. This meant I could lay the liner (with an underliner) on a layer of building sand laid directly on the pool subsoil floor. If it had been a waterlogged site then the floor would have had to be concreted to stop ground water coming up and 'floating' the liner in a completed pool.

▲ 2 Laying out the fleece underlining. The pieces were joined together by heating the edges with a blowlamp.

This concrete box approach is used by some professional installers, as there is very little chance of it going wrong – but at a cost: a lot of resources and a lot of cash, £50k to £60k for a natural swimming pool is not uncommon.

Outside the swimming zone the pool floor was formed into a giant basin shape and compacted with a petrol engine driven Wacker plate from a local tool hire company.

Lining the Pool

My greatest expense was the pool liner. It was also one of the hardest purchasing decisions. There is so much conflicting advice around, a lot of it from manufacturers claiming superiority of their product. Pond liners come in various thicknesses and materials. A thicker liner is obviously stronger and more expensive but it is also heavier to manipulate. I opted for 0.75mm EPDM from Flexible Lining Products. Although I think this was more a random choice born from a frustration of indecision, it seems to hold water, however, so not a bad decision in the end.

The liner was going to be buried in shingle contained in a 0.3m (1ft) deep ditch around the perimeter of the pool, and formed into a curtain drain. This keeps water run-off from the field from entering the pool and introducing nutrients, which would encourage algae. So, taking this into account, I needed a liner 26 x 20m. It cost £2,300 – the most expensive bit of plastic I have ever bought.

Underlining

A fleece underliner, from the same supplier, was laid in strips over the whole floor and walls of the pool (2). As part of some film research I was doing, I had just been to see The Swimming Pond Company install a pond in Suffolk and I picked up a vital tip. The fleece underliner, supplied in a roll, is laid in strips. It needs to be stuck to the next

▲ 3 The liner finally unfolded into place.

▲ 4 Overliner in place and planting areas defined by tyre walls.

strip to form a blanket over the whole pool area. Strips can be bonded to each other with a blowlamp. A very quick sweep of the flame along the edge melts a few fibres, so pressing this onto the edge of the next sheet makes them stick together.

The Liner

When the underliner was complete, the liner was brought next to the pond basin by a friendly farmer with a Teleporter (a tractor with a large retractable hydraulic arm) and placed onto a small scaffold rig. The roll was suspended on a scaffold pole threaded through the cardboard former the supplier had rolled the liner onto. Now it could be pulled and unrolled rather like a toilet roll, but bigger. The liner was 485kg and it bent the scaffold pole. Nonetheless, my partner and I managed to roll it out (3).

A 26m, half ton snake of liner folded like a concertina. We 'rippled' it along, inch by inch, with a fence post held between us and under the folded liner using a sort of peristaltic motion, rather like the pump in a dialysis machine. We then unfolded the liner and wafted the edges up and down to get air under to help it 'fly' over the whole area. I recommend you get as many friends as possible to help. It still would have been hard work even if there were ten of us.

It was about this time when I spoke to Michael Littlewood and he sent me his book, *Natural Swimming Pools, A Guide for Building*. It was great to have some real information at last.

I had made paper models of how the liner should be folded within the shape of my pool. This was very helpful because I knew what shape I was aiming for and where the big folds should come.

Overlining & Drainage

When the liner was in place and as many creases shuffled out of it as possible, a fleece overliner was laid on top. On top of this sand bags filled with a weak sand

▲ 5 Soil being added to the planting areas. The swimming zone is separated off by a low sandbag wall.

▲ 6 The finished natural swimming pool. The planting zones and swimming zone are clearly defined.

cement mix (10 parts sand, 1 cement) were placed immediately above the concrete wall defining the swimming zone (4). The wall was effectively continued up with more sandbags to a height of around 400mm. Each row of sandbags pushed back about 25mm compared to the row below, making the wall lean outwards against the ballast it has to retain.

Flexible drainage pipe was laid around the wall and then buried in shingle. The pipe terminated by emerging through the sandbag wall and into what would be the swimming area. This ultimately helps the water circulate below the roots of the regeneration zone. If necessary, a solar powered pump can be fitted but my pool water, so far, is perfectly happy without any artificial circulation.

I put a geotextile membrane over the shingle and covered it with many tons of the stony sand that had been excavated from the hole to make the pool (5). This was all done by hand because machines would damage the various linings. Around the pool I put up a chestnut paling fence. This is for safety; keeping children or visitors from straying near the pool. It also helps as a windbreak while the newly planted bushes and trees are too small to contribute any resistance.

Filling the Pool Naturally

Then it was just a matter of letting the pool fill with rainwater. I pumped it from the water butts around the house as well. Even with this addition it still took about a year to fill up (over here in East Anglia we don't get that much rain), but it was well worth waiting for. If I had used tap water the pool could have been more prone to algae problems. This is because of the phosphorous that is added to mains water, which is effectively a fertiliser. Having said this, commercial installers use mains water, but their pools then rely on powerful circulation pumps and filters, including phosphorous filters to help remove the impurities in the water.

▲ 7 This underwater picture demonstrates the excellent water quality achieved.

▲ Much of the planting is beautiful as well as useful.

Planting Up

It was deeply satisfying eventually putting plants into the sand (6). I had to select them to be 'soft rooted'. I sought guidance from Michael Littlewood's book. And today, the iris and *Ranunculus*, lilies and curly pondweed are all doing their job wonderfully. Most of the pool and the bank I have just left for wild plants to colonise and the sandy banks are now home to some beautiful tiny native flowers, as well as my friend, couch grass, some of it growing below the water line. And, so far, I have not needed to artificially pump the water around at all. The plants and animals keep the water crystal clear (7). Chemical free!

A Meeting of Minds

While I was building this pool, I made a film for BBC East 'Inside Out', on natural swimming ponds, and I was privileged enough to meet professional pool builders and Peter Petrich himself. As well as the interview and filming, I had the opportunity to discuss at length some of my non-conventional ideas on natural pools. I thought he would dismiss them, but instead, he was very supportive. It was heartening. I also spoke with Michael Littlewood. He, like me, also believed that some commercial companies make their pools far more complicated than they need to be.

Natural Pool Benefits

Building my own natural swimming pool has been my most rewarding experiment. Three years in the making, the ecosystem is stabilising and the water is sparkling clear. Just like that couch grassed water butt. I even became fitter than I have ever been with all that digging. And those couple of years of hard work ripple away with every splash of a bathing swallow, and each sight of a kingfisher hunting for water beetles. And, of course, there is the joy of swimming in soft rainwater! Your skin feels soft and healthy and your eyes don't sting with chlorine. One day I

▲ Naturally clean and clear water is soft and healthy for skin.

think we will look back and wonder how we ever thought it was reasonable to let our children swim in anything other than natural water.

Resources

Before building a pond, seek planning advice from your local planning authority on whether you need to apply for planning permission.

For excellent guides to creating ponds, see: **freshwaterhabitats.org.uk/habitats/pond/create-pond/make-garden-pond**

Peter Petrich's company website: **www.biotop-natural-pool.com**

Where David sourced his liner: **www.flexiblelining.co.uk**

The Swimming Pond Company: **www.theswimmingpondcompany.co.uk**

Costings

2,300	liner
700	underliner/over
1,000	diggers
500	shingle
400	block
200	cement
900	other stuff
£6,000	**Total**

Recommended Reading

Most of these titles are available to purchase through *Permaculture* magazine's online shop: **www.green-shopping.co.uk**

Building a Low Impact Roundhouse
Tony Wrench
Permanent Publications, 2014 (4th ed.)

Building a Wood-Fired Oven: For Bread and Pizza
Tom Jaine
Prospect Books, 2011

Building with Cob: A Step by Step Guide
Adam Weismann & Katy Bryce
Green Books, 2006

Clay & Lime Renders, Plasters & Paints
Adam Weismann & Katy Bryce
Green Books, 2015 (2nd ed.)

Coppicing and Coppice Crafts: A Comprehensive Guide
Rebecca Oaks and Edward Mills
The Crowood Press, 2015

Do It Yourself 12 Volt Solar Power
Michel Daniek
Permanent Publications, 2016 (3rd ed.)

From the Wood-Fired Oven: New and Traditional Techniques for Cooking and Baking with Fire
Richard Miscovich
Chelsea Green Publishing, 2013

Greenwood Crafts: A Comprehensive Guide
Edward Mills and Rebecca Oaks
The Crowood Press, 2012

The Hand Sculpted House: A Practical and Philosophical Guide to Building a Cob Cottage
Ianto Evans, Michael G. Smith, Linda Smiley, Deanne Bednar
Chelsea Green Publishing, 2002

How to Build a Natural Swimming Pool: The Complete Guide to Healthy Swimming at Home
Wolfram Kircher and Andreas Thon
Filbert Press, 2016

Living Wood: From Buying a Woodland to Making a Chair
Mike Abbott
Living Wood Books, 2002

Natural Swimming Pools: A Guide to Designing & Building Your Own (DVD)
David Pagan Butler
Permanent Publications, 2010

The Rocket Mass Heater Builder's Guide: Complete Step-by-Step Construction, Maintenance and Troubleshooting
Erica and Ernie Wisner
New Society Publishers, 2016

Roundwood Timber Framing: Building Naturally Using Local Resources (book and DVD)
Ben Law
Permanent Publications, 2010

The Solar Food Dryer: How to Make Your Own Low-Cost, High Performance, Sun-Powered Food Dehydrator
Eden Fobor
New Society Publishers, 2006

Wood Pallet Projects: Cool and Easy to Make Projects for the Home and Garden
Chris Gleason
Fox Chapel Publishing, 2013

Woodland Craft
Ben Law
GMC and Permanent Publications, 2015

Woodland Way: A Permaculture Approach to Sustainable Woodland Management
Ben Law
Permanent Publications, 2013 (2nd ed.)

Your Brick Oven: Building It and Baking in It
Russell Jeavons
Grub Street Publishing, 2004

Conversions

1cm/10mm = 0.394in
1in = 2.540cm/25.4mm

1m = 3.370ft
1ft = 0.305m

1km = 0.621mi
1mi = 1.609km

1ha = 2.471ac
1ac = 0.405ha

1g = 0.035oz
1oz = 28.350g

1kg = 2.205lb
1lb = 0.454kg

1 tonne = 0.984 ton
1 ton = 1.016 tonne

1ml = 0.035floz
1floz = 28.413ml

1l = 1.760 pints
1 pint = 0.568l

^{o}C to ^{o}F
double figure and add 30